Dedication

I would like to dedicate this work to my wife and children, for all of the years of wonderful questions they have asked have made me keep looking for the answers.

Know that science never has and never will prove anything! The goal of science is to provide evidence that supports some claim. That does not mean it is right nor does that mean it is wrong. Only the best evidence we can provide at this time. Question everything and never stop looking for new evidence that supports a new view. That is the one thing that keeps science dynamic. And yes, *Lampyridae* is a family of insects in the beetle order Coleoptera. They are winged beetles, and commonly called fireflies or "lightning bugs" for their use of bioluminescence to attract mates. Tomorrow they may be classified as something different because some new piece of evidence is found to show they are more closely related to the "Flood Monkey's" of Cedar Rapids!

That being said:

> If knowledge is power . . . and power is strength . . . and strength is the opposite of weakness . . . and weakness breeds fear . . . and fear nurtures contempt . . . and contempt fuels hate . . . and hate begets war . . . therefore: knowledge brings Peace. So, by buying this book and taking my class we will bring about world peace! Awesome! And thank you for stopping by.

FEAR and Loathing

An Introduction to Plant Biology

Thomas V. Vogel

Western Illinois University

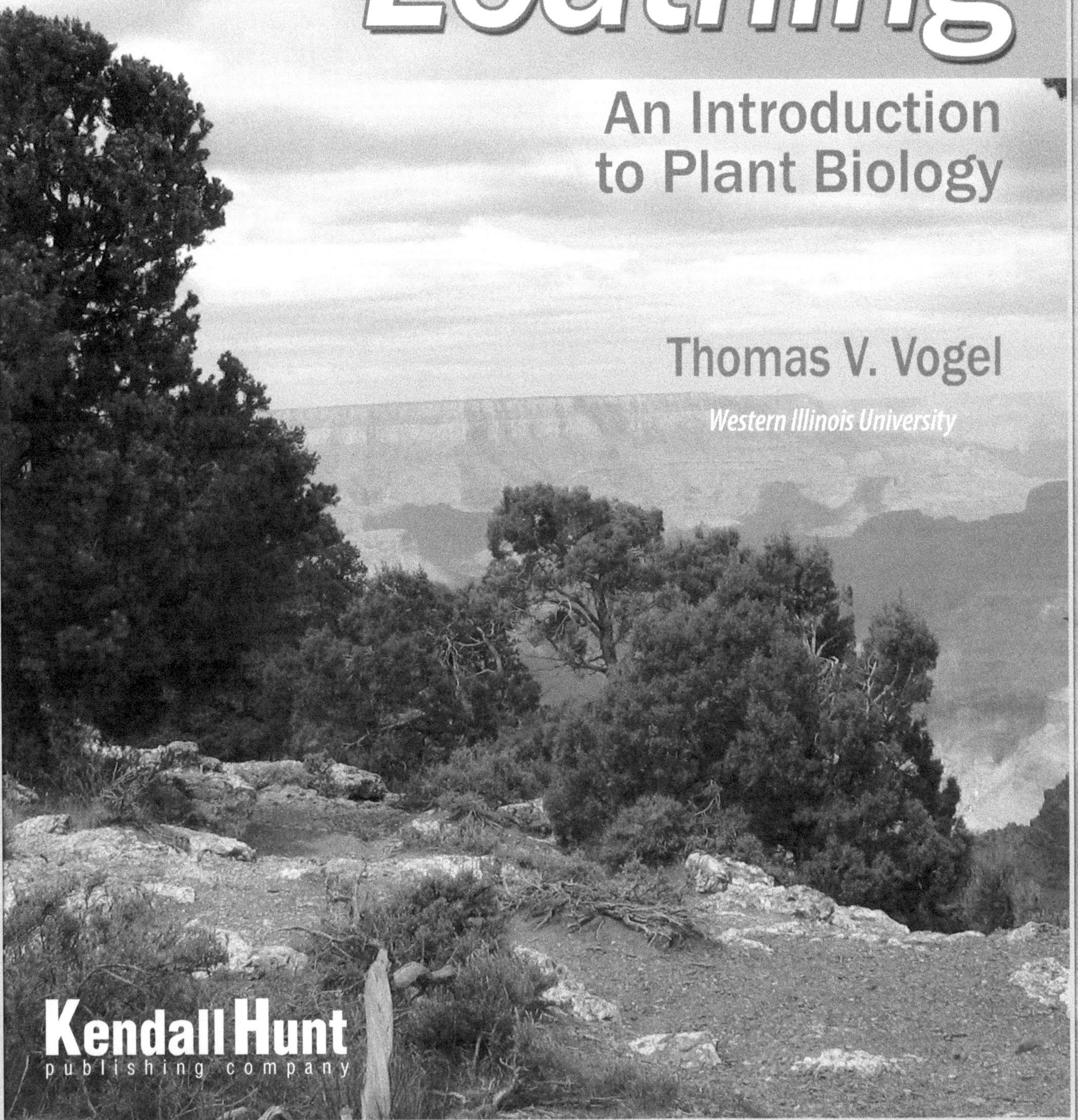

Kendall Hunt
publishing company

Cover image © Thomas V. Vogel

Kendall Hunt
publishing company

www.kendallhunt.com
Send all inquiries to:
4050 Westmark Drive
Dubuque, IA 52004-1840

ISBN: 978-1-5249-6891-5

Published in the United States of America

CONTENTS

PREFACE

This course is divided into six units. Each unit will have an exam that will summarize the material and allow for group study and provide opportunity for cooperative learning. All exams are in multiple-choice format and will be found online.

There will be a cumulative final exam. The lecture portion of this class will make up 75% of the total grade and the remaining 25% will come from Lab. I will not be recording attendance, although I do expect you to attend Lecture whenever you can. I would estimate that roughly 70% of the material I expect you to know for each exam is in this text, with 25% coming from the stories and movies shown during class. The remaining 5% will come from outside sources and from the use of your very capable minds.

INTRODUCTION

Don't panic! Trust me when I tell you it will be fine. You are not the first class of students that will experience the angst of Intro Biology and I am certain you will not be the last. As you huddle in the comfort of the cinderblock room that will be "home" for the next four years (yes, Virginia, you can graduate in four years, although it did take me nine years for my Bachelor of Science!), remember to pace yourself. There is time enough for the full college experience. I would hate for you to miss anything that will make for a colorful story for your children and grandchildren. As this is an introductory course, we are forced to include a phenomenal amount of material. Because we cover so much, we will only be touching on each of the topics for a short time before we find ourselves moving on. Please feel free to ask as many questions as you wish. I will not have all of the answers. I do promise to do my best to find the most complete and timely answers to anything you ask—safe from prejudice, blame, fear, or failure. The more questions you ask, the more time I can devote to teaching and less time lecturing. That's right, my standing in front of you three days a week and talking is not the key to learning. Learning is a permanent change in behavior due to experience. How many times did your parents warn you "Don't touch that, it's hot," only to wait until no one was watching you so you could finally touch it. Don't be ashamed. Everybody had to do the same thing to learn what "hot" means. I do not see my role as your teacher—life is your teacher. I am just your storyteller. Hopefully, something you have heard in class will be there in your subconscious. When life sends you an experience, you will remember the story from the lecture and have that light bulb moment we live for. Learning is in the doing—lecturing to you is my way of giving you a preview, so when the experience comes, it is easier for you to make sense of it and learn from it. Or in the case of my less traditional students who may have witnessed many of these events already, this will help to clarify what happened or why. To my family and all overeager to learn: Thank you and hang on; we won't be back this way again!

UNIT ONE

UNIT I

Nature of Science
Cells
Angiosperm Anatomy

Nature of Science
Deductive versus Inductive

- **Deductive Reasoning**—making individual decisions by applying generally accepted principles
- Typical decision-making process

Nature of Science
Deductive versus Inductive

- **Inductive Reasoning** —way of discovering general principles by careful examination of specific cases
- Became popular about 400 years ago (in Europe)
- Develop a hypothesis about how the world works
- Perform experiments to support the hypothesis

Scientific Investigation
Controlled Experiment

- To evaluate alternative hypothesis about one variable
- All other variables must be kept constant
- This eliminates misleading or confusing results
  - Two experiments are done at the same time
    - This gives the validity to the conclusion
- Helps to reduce Experimental Error
- **Control Group**—does not get the Experimental Factor
- **Experimental Group**—exactly the same as the Control Group—plus it gets the Experimental Factor

The Heterotroph Hypothesis

- The protocell was most likely heterotrophic (takes in preformed food for energy)
- Lived in the primitive ocean where there was a ready supply of small organic molecules, including adenosine triphosphate (ATP)
- Probably had a limited ability to break down complex molecules for energy
- Since there was no free oxygen, it must have been a form of fermentation

The True Cell

- A membrane-bound structure
- Can carry on protein synthesis that can produce the enzymes needed for DNA replication to take place
- Capable of reproducing itself
- DNA-RNA-Protein is the standard accepted relationship today

Prokaryotic Cell

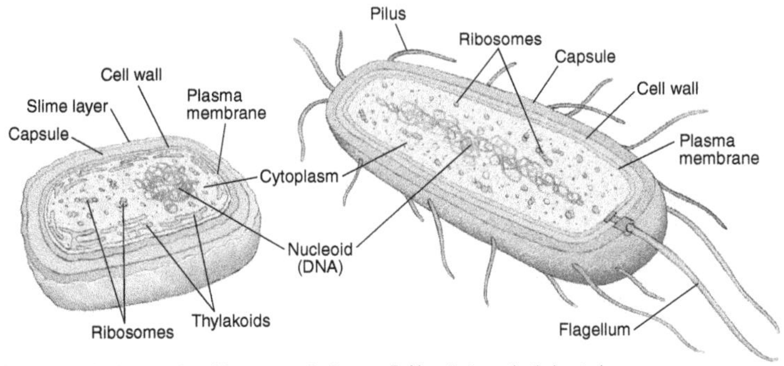

A. Photosynthetic bacterium (blue-green alga) B. Nonphotosynthetic bacterium

Origin of Life
Evolution of Eukaryotic Cells

- **Endosymbiotic Hypothesis**
- Prokaryotic cells enter a symbiotic relationship with other prokaryotic cells
- Small cells enter a large cell through endocytosis
- This creates a large cell with small specialized cells (organelles) inside
- When the large cell divides, so do the small cells
- This may have produced the first eukaryotic cell

Multicellularity

- Some unicellular organisms began living in association with others
- Safety in numbers
- Take advantage of special abilities of some organisms
- Unicellular → Colonies → Multicellular

Eukaryotic Cell

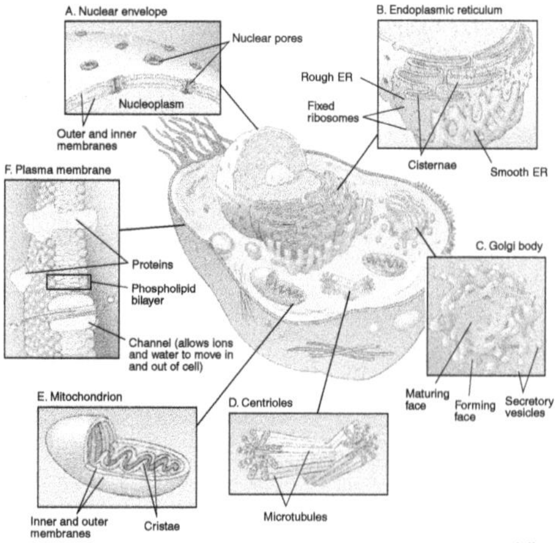

Cell Size

- **Surface Area to Volume Ratio**—nutrients enter and wastes exit through the membrane on the surface
- **Volume**—represents the needs of all cells
- **Surface Area** —represents the cells ability to get material in and out of the cell

*As the volume increase through growth, the surface area to volume ratio decreases

Cell Structure and Function

Cell Membrane

- Thin, nearly invisible structure that surrounds the cytoplasm
- Connects the endoplasmic reticulum with the nuclear membrane

Cell Membrane

Made up of two parts

A. **Phospholipids**—with tails in the diagram

B. **Protein**—center of the diagram

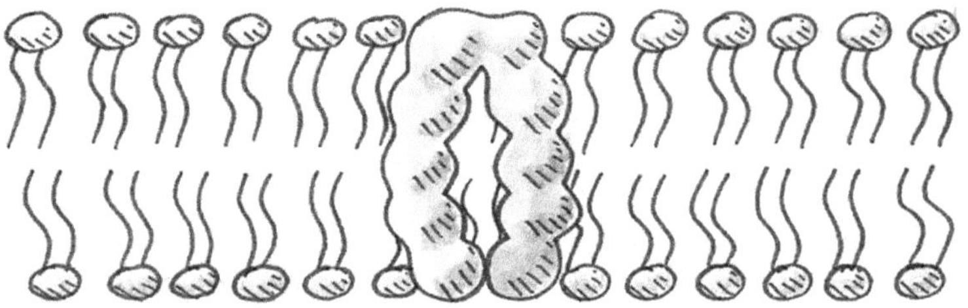

image courtesy of Kevin Mercer

Cell Membrane

Three types of protein molecules found in the membrane

1. **Transport**—regulate transport and diffusion
   - **Carrier Proteins**—move molecules one at a time through the lipid layer; they do not extend through the membrane

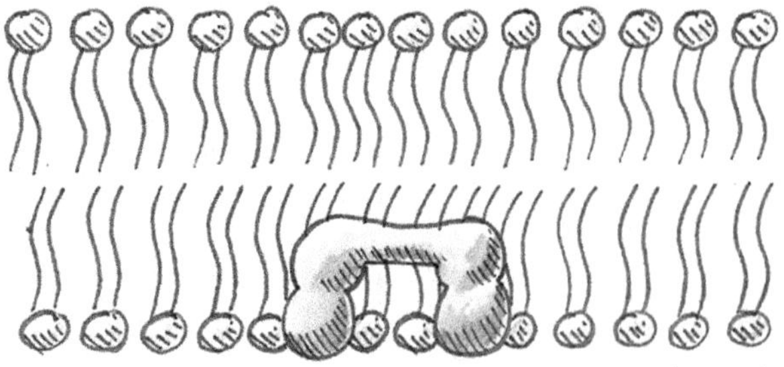

image courtesy of Kevin Mercer

Cell Membrane

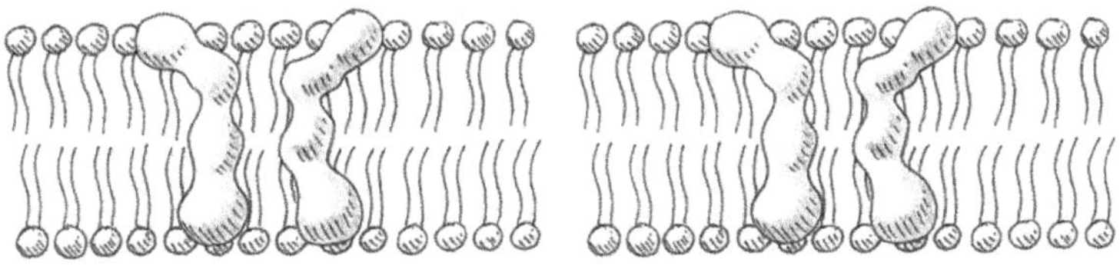

- **Channel Proteins**—extend through the lipid layer and allow for the random diffusion of molecules through the membrane

image courtesy of Kevin Mercer

Cell Membrane

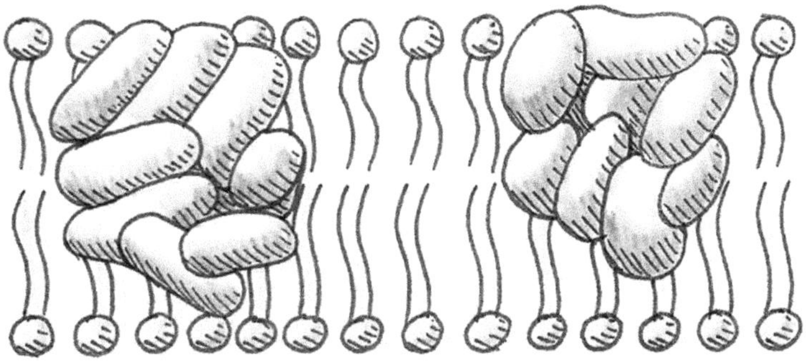

2. **Marker Proteins**—identify cells to other cells (immune system)

image courtesy of Kevin Mercer

Cell Membrane

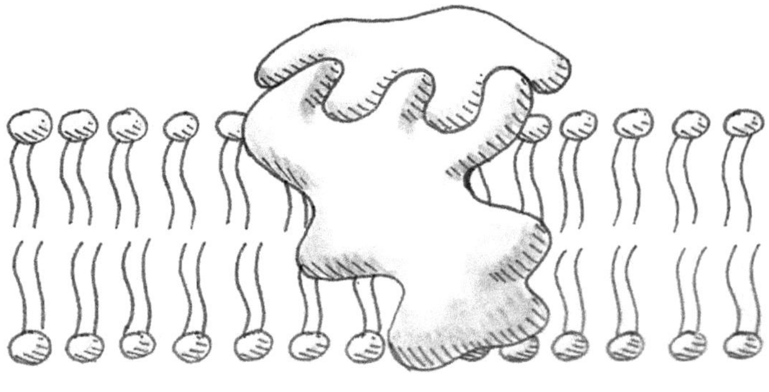

3. **Receptors**—allow the cell to receive instructions

image courtesy of Kevin Mercer

Cell Membrane Structure and Function

- Regulates the passage of molecules in and out of the cell
- **Passive Transport**
  - **Diffusion**—needs a concentration gradient only, moves to lower concentration
  - **Facilitated Diffusion**—needs a carrier molecule and concentration gradient, moves toward lower concentration

Passive Transport Diffusion

- Movement of molecules from an area of higher concentration to an area of lower concentration (down the concentration gradient) until an equilibrium is achieved
- KEY POINTS:
  - Always down the concentration gradient
  - Requires no chemical energy
  - Continues until equilibrium is reached

OSMOSIS
= the diffusion of water in and out of cells

- A differentially permeable membrane separates the cell and its environment
- A concentration gradient exists betweens the cell and its environment
- The differentially permeable membrane allows for the passage of a Solvent (water), but not Solute
- Process will continue until equilibrium is reached

Osmosis

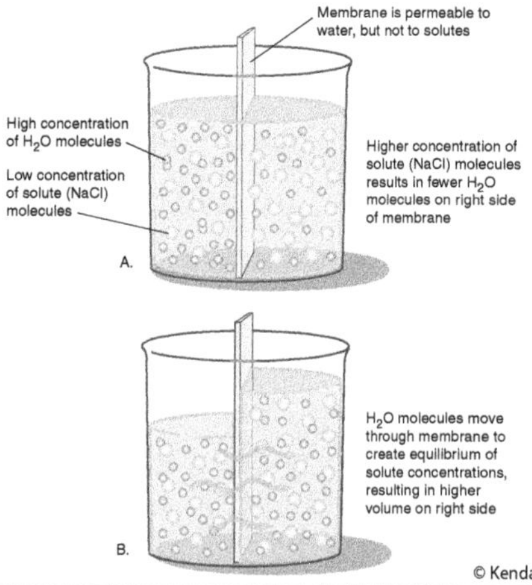

Osmotic Pressure

- The pressure that develops in a system due to osmosis
- The greater the osmotic pressure, the greater the likelihood that water will diffuse in that direction

Tonicity

Isotonic Solution

- The solute concentration is the same on both sides of the membrane; therefore, there is no net gain or loss of water
- Examples:
  - A 0.9% solution of NaCl is known to be isotonic to red blood cells
  - Intravenous injections, medically administered

Tonicity

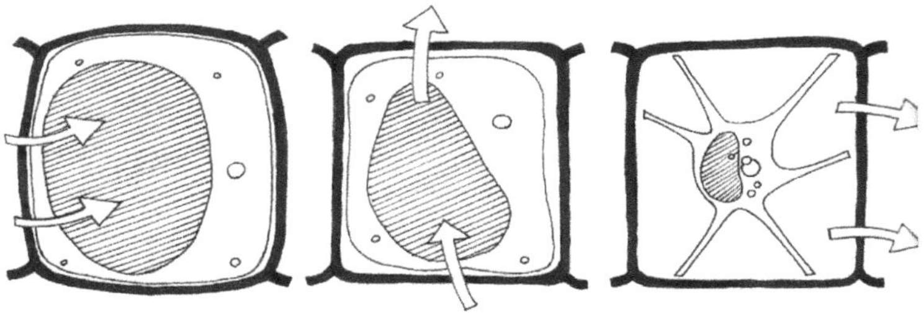

image courtesy of Kevin Mercer

Hypotonic Solutions

- The prefix hypo—means less than, and refers to a solution with a lower percentage of solute (more water) than the cell
  - A cell placed in a hypotonic solution will show a net increase in volume
  - There will be a net flow into the cell
- Examples:
  - Any solution with a concentration lower than 0.9% is hypotonic to red blood cells
  - Animal blood cells placed in such a solution will swell and burst (Hemolysis)
  - Animal cells placed in such a solution will swell and burst (Plasmolysis)

Hypertonic Solution

- The prefix hyper—means more than, and refers to a solution with more solute (less water) than the cell
  - A cell placed in a hypertonic solution will show a net decrease in volume
  - There will be a net flow of water out of the cell
- Examples:
  - Any solution with a concentration higher than 0.9% is hypertonic to red blood cells
  - Animals placed in such a solution will shrink (Crenation)
  - Meats preserved by salting, the salt removes water from the meat, thereby killing the bacteria

Hypo-Hypertonicity

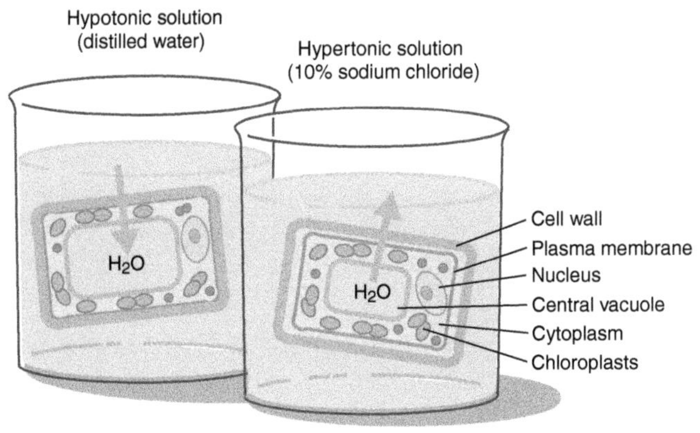

A. Net flow of water into cell B. Net flow of water out of cell

Transport by Carrier Proteins

Active Transport

- Carrier molecules are specific
- Explain the passage of amino acids and glucose across the plasma membrane
- Requires chemical energy
- Can occur as often as 100 times a second

Eukarya (Eukaryotes)

- Appear about 1.5–2.0 bya
- Include the following four Kingdoms:
  - Fungi
  - Plants
  - Animals
  - Protists

Plant Structure, Growth, and Development

- When plants began living on land (~420 mya), distinct tissues and organs evolved
- All flowering plants have leaves, stems, and roots
- Structures may be highly modified and/or reduced

- No species has completely lost roots, stems, or leaves
- Division Magnoliophyta = informally "**angiosperms**"
  - **Eudicots** (broadleaf plants)
  - **Monocots** (narrow-leaved)
- The **primary (1°) plant body** is herbaceous
- The **secondary (2°) plant body** is "woody"
- **Herb:** plant never becomes woody; lives < 1 year

Basic Types of Cells and Tissues

- Stems, leaves, and roots all share a basic simple organization
- All plant cells belong to three classes based on the nature of cell walls and cell functions:
  - Parenchyma
  - Collenchyma
  - Sclerenchyma

Parenchyma

- **Parenchyma Cells:** thin 1° walls; "workhorse cells"
- **Parenchyma Tissue:** mass of parenchyma cells

- Some parenchyma cells function when dead at maturity (xylem)
- Parenchyma cells relatively "cheap" to build (thin 1°walls)
- Most leaves: flexible, composed almost entirely of parenchyma
- Glandular parenchyma cells secrete:
  - nectar, fragrances, mucilage, resins, and oils

Collenchyma

- **Collenchyma Cells**—thin 1° wall, thickened in some areas (corners)
  - allow flexible support to growing regions of shoot
  - located:
    - beneath epidermis
    - supporting vascular bundles in growing shoot tips (celery stalk "strings")
  - in shoot (stem + leaves)

Sclerenchyma

- 1° wall and thick 2° wall (usually lignified)
  - elastic; deformed but return to original size and shape
  - Protective seed cover, fibers in stem for support (fibers), etc.
  - Support plant thallus by its strength
  - 2° cell walls prevent protoplast from expanding when water absorbed
  - Dead at maturity/when functional/once 2° cell wall produced

- Sclerenchyma Types:
  - Fibers: long and flexible
  - Sclereids: short, isodiametric (cuboidal), inflexible, and brittle
  - Sclereids and fibers differentiate from parenchyma cells

External Organization of Stems

- **Shoot:** stem + leaves/aboveground portion of thallus
- **Nodes:** where leaves attached
- **Internodes:** regions between nodes

- Stem area just above point where leaf attaches: **leaf axil**
- Contains:
  - **axillary bud:** mini shoot w/dormant apical meristem, at leaf base
  - several young leaves
- At extreme tip, each stem **terminal bud**

- **Differentiation:** process whereby unspecialized cells/tissue becomes specialized in form or function
- All cells in thallus contain same genes/DNA, just "turn-off" genes not needed
- Unspecialized cells can divide and produce cells that specialize so plant can grow and develop

Primary Apical Meristems:

- regions that generate new 1° tissues at root/shoot tips
  - elongate stem/root
  - give rise to all tissue types:
    1. **Epidermal:** "skin" of plant, waterproofing, protection, etc.
    2. **Vascular:** conduction of H_2O and substances throughout thallus (xylem and phloem)
    3. **Ground Tissue:** located between epidermis and vascular tissues; mostly parenchyma

- Plants grow by ⇑ cell number and/or ⇑ cell size
- Cells can easily and cheaply enlarge by H_2O uptake
- **Secondary (2°) Meristems**: meristem regions that give rise to 2° growth by producing "wood" and bark
  - cause ⇑ in girth, support perennial plants' growth
  - provide us with wood

Leaf Structure

Gross Morphology

- **Blade**—the flat portion of the leaf that functions in photosynthesis
- **Petiole**—attaches the blade to the stem
- **Stipules**—**leaflike** growths that may occur at the base of the petioles

Leaf Structure

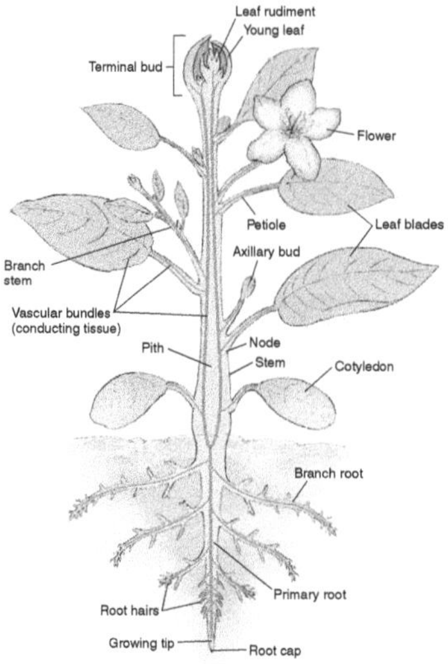

© Kendall Hunt Publishing Company

Leaf Structure

Gross Morphology

- **Simple leaves**—have only one blade per petiole
- **Compound leaves**—blades are divided into leaflets
- **Lateral Buds**—develop in the leaf axial (between petiole and stem)
  - Never develop at the base of leaflets (used to differentiate between a leaf and a leaflet)

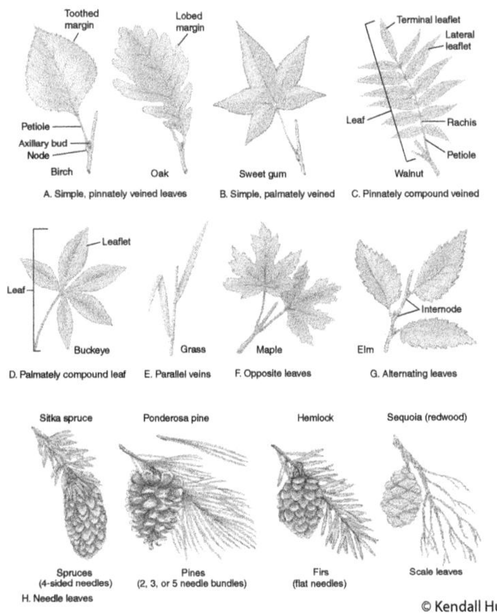

Eudicots versus Monocots

- Monocots
  - Parallel venation
  - Lack a petiole
  - Mesophyll is not differentiated into spongy and palisade layers
  - Leaves (in cross section) showing evenly spaced vascular bundles

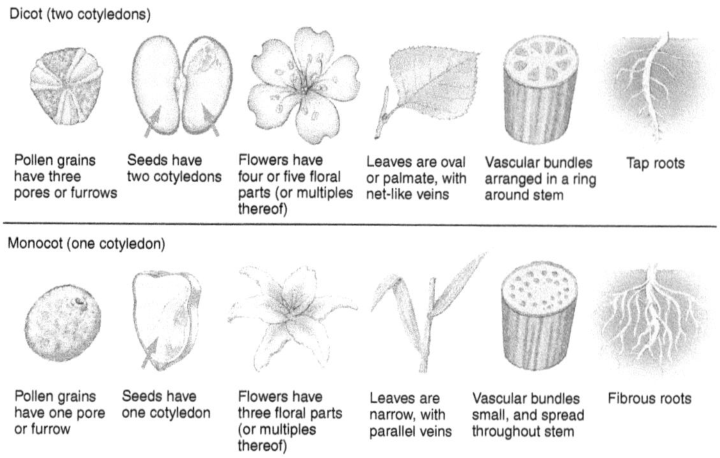

Eudicots versus Monocots

- Eudicots
  - Netted venation
  - Petiole present
  - Mesophyll is differentiated
  - Leaves (in cross section) showing randomly spaced vascular bundles

Stems and Material Transport

Stem Basics

- **Stem**: region between leaves and roots of vascular plants
- **Functions**:
  - elevate leaves
  - transport water, sugars, hormones, molecules between leaves and roots
  - produce flowers and fruits
  - (maybe) storage (examples: potatoes, etc.)
  - produce hormones for thallus

- Modifications increase stem's ability to survive, spread, or store nutrients
- Although axil of every leaf contains bud, only few buds develop into branch; others remain dormant or produce flowers
- Most plants, axillary buds not needed if apical meristem healthy; if damaged/killed, axillary buds become active/develop

Internal Organization of Stems: Arrangement of 1° Tissues

- **Epidermis** = outermost surface of herbaceous stem; single layer parenchyma cells
  - All exchange of material between plant and environment occurs via epidermis
- **Functions**: protection and prevent water loss
- Outer tangential walls coated w/waterproof cutin; layer called **cuticle**

- Cuticle prevents gas exchange, so accomplished by **stomata**
- Stomata: pairs of **guard cells** surrounding **stomatal pore**
- Guard cells swell when absorb water
    - Guard cells swell when absorb water
        - pore opens, allow exchange CO_2 and O_2
    - Water lost through stomata
        - = Balance open/closed with water needs

Stem

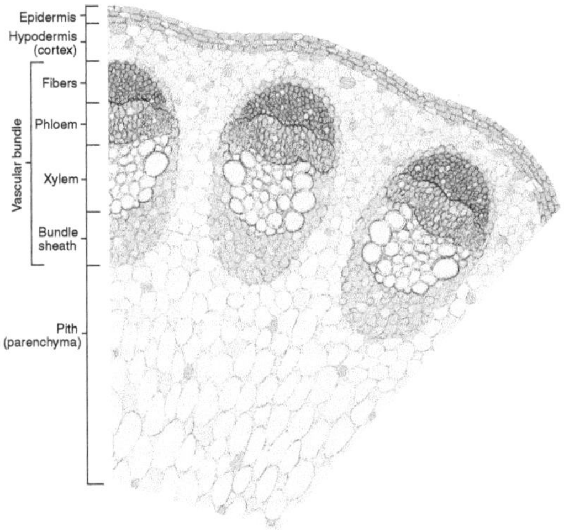

- Some epidermal cells elongate outward and become **trichomes** (hairs)
  - protect against animal attack/water loss/too intense sunlight
  - May produce oils and antiherbivore/microbial chemicals (protection)
  - **Cortex** = interior to epidermis; composed of photosynthesis (psn) parenchyma and sometimes collenchyma

- **Vascular Tissues**
  - Two types **vascular tissues:**

    1. Xylem: conducts water and minerals up thallus

    2. Phloem: distributes sugars and minerals up/down
  - Xylem dead and hollow at maturity, 2° cell wall
  - Phloem alive at maturity; transports dissolved materials, 1° cell wall only

- Xylem
  - = **tracheids** and/or **vessel elements**
    - types of sclerenchyma and **tracheary elements**
  - Simplest form, 2° cell wall organized as set of **annular thickenings** on inside 1° cell wall
  - **Perforations** form between vertically stacked vessel elements
  - Stack of vessel elements: **vessel**
  - Angiosperms have tracheids and vessel elements
  - Gymnosperms have only tracheids

- Phloem
  - Two types of conducting cells:
    - **sieve cells**
    - **sieve tube members**
    - Term "**sieve element**" refers to either:
  - Live parenchyma cells at maturity, 1° cell wall only
  - Plasmodesmata enlarge to become **sieve pores**; aggregate in **sieve areas**

- **Sieve tube members:**
  - wider sieve areas on each end wall, called **sieve plates**
  - align vertically, forming **sieve tube**

- Nuclei of sieve elements degenerate
  - associated with neighboring cells that exert nuclear control
  - end walls develop open areas: sieve areas
  - sieve tube members associated with **companion cells**

- Sieve tube members controlled by **companion cells**
  - load sugars
  - smaller than accompanying conducting cell
  - prominent nucleus and dense cytoplasm filled w/ribosomes
- **Vascular Bundles**
  - Xylem and phloem occur together in **vascular bundles** interior to cortex

- **Eudicots**: vascular bundles arranged in <u>one ring surrounding</u> **<u>pith</u>**: central region of parenchyma similar to cortex
- **Monocots**: <u>distributed throughout inner part of stem</u>
- All vascular bundles **collateral**: each contains both xylem and phloem strands running parallel to each other

Eudicot Stem

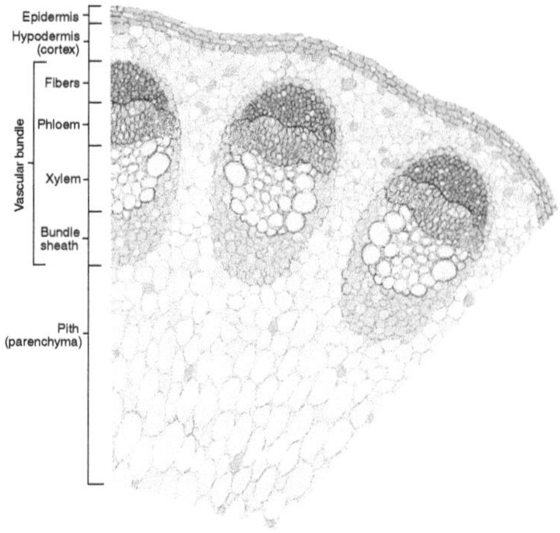

- Xylem of vascular bundle: **primary xylem**
  - part of 1° plant body
  - large proportion xylem parenchyma and xylem fibers (strength)
- Phloem of vascular bundle: **primary phloem**
  - storage parenchyma and phloem fibers or sclerids associated with sieve elements and associated cells

Monocot Stem

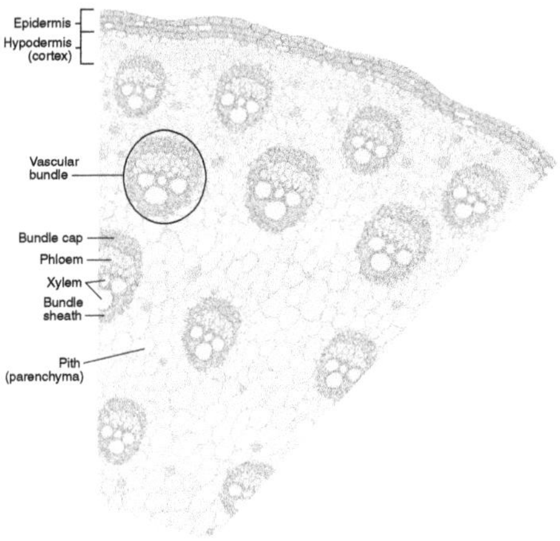

Stem Growth and Differentiation

- Stems grow longer by creating new cells at tips, in **apical meristems**
    - Cells continuously divide/remain mitotic
- As smaller daughter cells expand and grow, push meristem outward

- Region below apical meristem: **subapical meristem**
  - cells also divide and grow, slower rate
  - some cells here stop dividing and differentiate into **protoxylem** cells

- Cells exterior to protoxylem begin to differentiate next
  - larger because of longer expansion
  - are **metaxylem**
- Similar process occurs outer part of each vascular bundle
  - Exterior cells mature: **protophloem**
  - Cells closest to metaxylem become **metaphloem**
- Other tissues also differentiate in subapical region: epidermis, pith, cortex

- **Protoderm:** epidermal cells in early stages of differentiation
- **Provascular tissues:** young xylem and phloem cells:
  - **Ground meristem:** equivalent stages of pith and cortex
  - **1° tissues** produced by apical meristems
  - **1° growth:** growth and tissue formation results from apical meristem

Transport Basics

- Substances must be moved throughout thallus
- **Short-distance transport** = transfer of basic nutrients, cell to cell
- **Long-distance transport**, usually via xylem and phloem, adaptive for land plants

- Vascular tissues selectively advantageous for shoots to grow upright, allow psn
- Move substances to from where produced → where needed
- Almost everything transported dissolved in water, so ability to move H_2O essential
- **Molecular pumps** use ATP to force molecules across membrane, even against concentration gradient

H_2O Potential

- H_2O has free energy, a capacity to do work, called H_2O **potential** (ψ, "psi")
- Pure water = much potential
- More solutes dissolved in H_2O = ↓ potential
- H_2O moves from regions where there is "more H_2O" → "less H_2O"

Cells and H_2O Movement

- If plant in dry/"overly fertilized" soil = loses too much $H_2O \rightarrow$ wilting or death
  - **Incipient plasmolysis** = point at which protoplast has lost enough H_2O to pull slightly away from wall (stressed)
  - If lose more H_2O, protoplast pulls completely away from wall and shrinks $\rightarrow$ **plasmolyzed**, dies

Cell
Plasmolysis

Short-Distance Intercellular Transport

Plant cells pass some materials short-distance through:

1. Plasmodesmata
2. Diffusion
3. Molecular pumps in membrane
4. Vesicle transport (intracellular/intercellular)
5. Apoplastic pathway

Long-Distance Transport: Phloem

- Best model for H_2O and nutrients movement: **pressure flow hypothesis**
  - substance moved from "source" to "sink"
    - at "source," sugars actively transported into sieve elements via membrane pumps
    - source: where something is produced or stored

- Sink: region where substance needed
- At "Sink," sugars actively unloaded from sieve elements into surrounding cells (growing/developing structures)
- H_2O moved to nearby xylem

Long-Distance Transport: Xylem

- H_2O is heavy
  - Lifting to great heights would take a great deal of energy.
- **Cohesion-Tension Hypothesis**: most widely accepted model
- H_2O unavoidably escapes through stomata when open = **(transstomatal) transpiration**

Cavitation

- **Cavitation:** situation where hydrogen bonds between H_2O breaks and water column in xylem breaks, leaving an open space (embolism)
- Tracheid or vessel never conducts H_2O again

- Adhesion between H_2O and cell wall helps prevent cavitation most effectively in narrower elements than wider ones
- Thus, in dry habitats, species possess more narrow tracheary elements than wider ones

- Eventually, all vessels and tracheids cavitate
  - After tracheary elements cavitate, surrounding parenchyma cells block them with tyloses or by secreting gums and resins, they then become part of heartwood in trees

Secondary Growth

- In herbs, only 1° plant body develops
- In woody species, additional stem and root tissues produced from lateral meristems: vascular cambium, cork cambium
- To allow ⇑ H_2O-conducting capacity, "epidermal" protection

- New tissues are:
  - wood (2° xylem)
  - bark (2° phloem, cork)
- **Secondary Tissues** = constitute plant's 2° body

- Must use energy and nutrient resources for winterizing thallus in temperate climates
- Wood and bark construction is metabolically expensive

- All woody trees and shrubs descended from one group of ancestral woody plants (~370 mya)
- Woody plant: combination of 1° and 2° tissues
  - Only parts of older stems and roots have 2° tissues
- True 2° growth occurs in:
  - Many eudicots, most basal angiosperms, all gymnosperms
  - **Not** in ferns or monocots

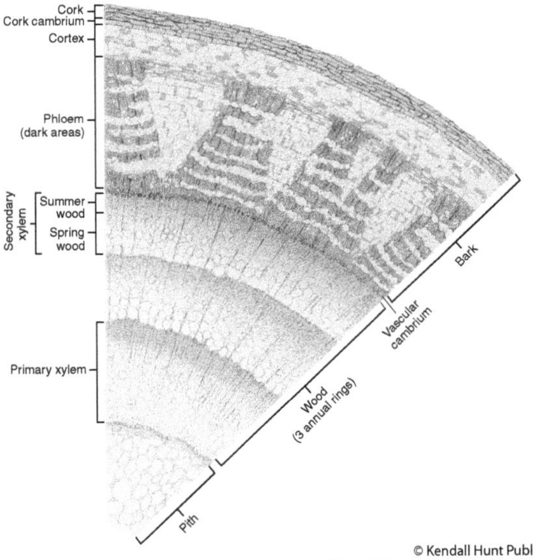

- **Hardwoods**—large amounts of fibers
- **Softwoods**—few or none

- Axial xylem parenchyma- temporary H_2O storage, short lateral movement
    - Conifers have little/no axial parenchyma
- Most gymnosperms contain only tracheids in axial systems
    - Fibers and parenchyma cells sparse/absent

Growth Rings

- Cells formed early in growing season: **early/spring wood**
    - High proportion of wide vessels or tracheids
- Late in growing season: **late/summer wood**
    - Fewer vessels or narrower, thick-walled tracheids
- Together, early wood and late wood make up 1-year growth - **annual (growth) ring**
- Patterns - very species specific

- **Heartwood and Sapwood**
  - Center of log (**heartwood**) almost always darker, drier, and more fragrant than outer functional wood (**sapwood**)

- Wood parenchyma cells produce compounds that inhibit growth of bacteria and fungi:
  - Large quantities of phenolic compounds
  - Lignin
  - Other dark, aromatic substances
- Once all parenchyma cells in area die, decay-resistant heartwood has formed

Secondary (2°) Phloem

- Axial system responsible for conduction up and down stem/root:
  - sieve tube members and companion cells in Angios
  - sieve cells and albuminous cells in Gymnos
  - sieve tube members and companion cells in Angios
  - both have fibers and some nonconducting parenchyma present nearby
- Phloem rays formed just as xylem rays are:
  - parenchyma short-term storage cells and albuminous cells in Gymnos

- **Cork and Cork Cambium**
  - As 2° tissues added and pushed outward, tissues on periphery either:
    - either grow in circumference or torn apart

Outer Bark

- **Cork and Cork Cambium**
  - As 2° tissues added and pushed outward, tissues on periphery either:
    - either grow in circumference, or torn apart
  - As circumference ⇑ and older sieve elements die, some storage parenchyma cells become mitotic = new cambium within 2° phloem: **cork cambium (phellogen)**

- After each division, inner cell remains cork cambium
  - outer cell differentiates into **cork (phellem) cell**
- Cork cambium may produce one to two cells to inside; mature into layer of parenchyma called **phelloderm**
- Cork cells filled with waterproof suberin and die = waterproof and chemically inert

- All plant material exterior to periderm dies due to lack of H_2O and nutrients
- Layer of cork cells cannot expand much circumferentially
  - New cork cambium forms in younger 2° phloem closer to vascular cambium

- All tissues outside innermost cork cambium comprise: **outer bark**
- All 2° phloem between vascular cambium and innermost cork cambium: **inner bark**

Initiation of Cork Cambia

- First cork cambium may arise in:
  - Epidermis
  - Cortex
  - 1° phloem
  - 2° phloem
- Subsequent cork cambia form afterward, deep in 2° phloem
- First bark on young stems differs from bark formed when stem older

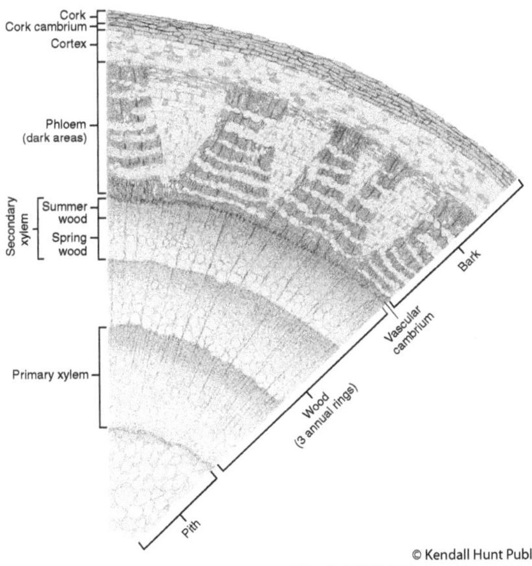

- 2° growth in monocots
  NOT true 2° growth
  - Region similar to vascular cambium arises outside outermost cortex vascular bundles
    - Only parenchyma cells produced

- Columns of some parenchyma cells produce narrow cells that differentiate into **2° vascular bundle**
- Fibers develop exterior to bundles that give these trees a "woody" character

UNIT TWO

UNIT II

Roots
Leaves
Photosynthesis
DNA and RNA

Roots and Plant Nutrition

- **Functions**:
  1. Anchoring plant firmly to substrate
     - Firm anchoring provides stability
  2. Absorb H_2O and minerals
     - soil: where the H_2O and minerals are
  3. Produce hormones
     - coordinate growth of thallus with shoot

- **Adventitious roots** arise <u>from stem</u>
  - root absorptive and transport capacities, or support stem
  - Adventitious/prop roots of Mangrove help it survive coastal tides
  - Prop roots help support shoot, provide stability in shifting/nonstable soils or in wetlands

External Root Structure: Organization of Root Systems

- Most seed plants have:
  - Single large taproot develops from **radicle** (embryonic root)
  - Numerous small **lateral roots** or **branch** roots develop
  - Lateral roots may also produce more lateral roots → highly branched root system

- Most monocots and some eudicots have a mass of many similarly sized roots
  - = **Fibrous root system**
    - Very effective at holding soil in place!

- **Root hair zone:** region in which many epidermal cells extend out as narrow trichomes; greatly ⇑ root's surface area!
  - <u>functional root absorptive area</u>
- New lateral roots emerge
  - behind root hair zone

Internal Root Structure

- **Root cap:** cells in layer closest to root apical meristem (meristematic cells) form files of cells that are pushed downward and continuously form root cap, dividing at slightly slower rate

- **Zone of elongation**
  - Cells expand greatly by absorbing H_2O
  - Protoderm, provascular tissue, and ground tissue begin to differentiate
    - No cells yet mature

- **Zone of maturation/root hair zone**
  - Root hairs grow outward
  - ⇑ absorption of H_2O and minerals
  - Cortex cells transfer minerals from epidermis to vascular tissue

- Innermost layer of cortical cells differentiates into cylinder: **endodermis**
  - Radial walls of endodermis imbedded with waterproof lignin + <u>suberin</u>: **casparian strip** controls what enters xylem

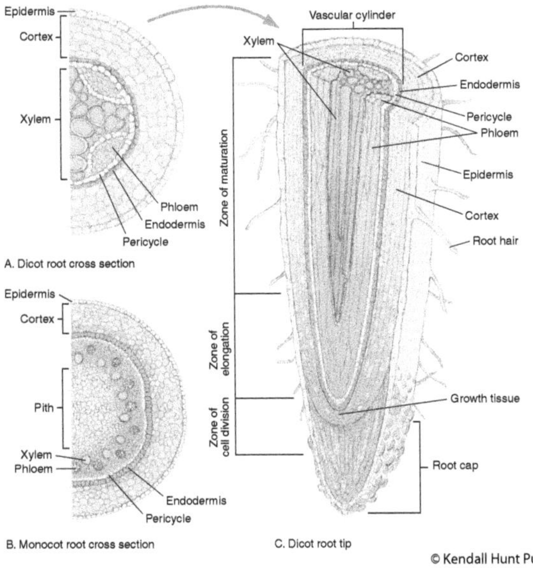

- Radial walls of endodermis imbedded with waterproof lignin + <u>suberin</u>: **casparian strip** controls what enters xylem
- Minerals can cross endodermis only if endodermal protoplasts absorb them from:
  - intercellular spaces of cortex apoplast
  - cortical cells
- Secrete them into vascular tissues

- **Root xs**: xylem (except in monocots) forms solid mass in center, surrounded by strands of phloem
- NO pith present, stele
- **Stele:** region in root center, composed of xylem and phloem

- Within xylem strands:
  - Inner wide cells: metaxylem
  - Outer narrow cell: protoxylem
- Within phloem strands:
  - Outer side: protophloem
  - Inner side: metaphloem
- **Pericycle:** irregular row of parenchyma cells between vascular tissue and endodermis
  - where lateral roots originate

- In mature portions of root:
  - Root hairs function only for several days
  - Root continuously produces new ones
  - In endodermis, more suberin + lignin laid over radial and tangential surfaces
  - Mature, watertight endodermis keeps water in stele of mature root

- Cortex and epidermis, aside from root hairs, may be superfluous
  - Shortly after root hairs die, underlying cortex and epidermis often die and shed from root
- Endodermis becomes root surface until bark can form
- Large fibrous roots of many monocots are:
  - strictly annual
  - completely replaced by more adventitious roots

Origin and Development of Lateral Roots

- A few pericycle cells become mitotic
  - new root apical develops and functions
  - grows through cortex and connected to stele

- Root hairs, especially with fungal partner, VERY efficient at acquiring elements from soil
- Molecular pumps actively transport elements into root hairs
- Root hairs selective in absorption
- Elements move apoplastically or symplastically to endoermis, then MUST enter from endodermis to stele

Other Root Cell Needs

- Root cells also need:
  - O_2 (diffusion from stem)
  - H_2O (from root hairs)
  - sugars (from stem via phloem)
  - get rid of CO_2 (diffusion into soil/stem)

- Components that are necessary for plant growth—**essential elements**
  - **Macro essential elements**—required in large quantities (N, P, K, Mg, etc.)
  - **Micro essential elements**—required in low concentrations (Cl, B, Fe, etc.)

Leaves: Photosynthesis and Transpiration

The Basics

- Leaves should be flat and thin for maximum absorption of light + CO_2
  - Most of the tissues must be alive and differentiated into chlorenchyma for photosynthesis (psn) and gas exchange
- Stems elevate leaves and conduct material to and from them, among other activities
- Almost all leaves contain only 1° tissues

External Structure of Foliage Leaves

- **Cuticle:** waxy-covering, secreted by epidermal cells; waterproof
- **Epidermis:** outer covering of thallus; continuous with entire thallus; NO chloroplasts!
  - ⇓ H_2O loss, ⇓ pathogen infection, allow gas exchange (O_2 and CO_2)
  - Special cells allow defense (trichomes), gas exchange (stomata and guard cells)

- Cuticle and epidermis are good/bad:
    - + waterproof thallus
    - + ⇑ pathogen resistance
    - – greatly interfere psn!
    - = plants evolved **stomata** to allow it to avoid limitations, yet survive

- Leaf epidermis is often "hairy"
- **Trichomes** affect leaf biology:
    - provide shade on leaf upper surface
    - prevent rapid air movement, ⇓ transpiration
    - make insect walking/chewing difficult
    - many glandular trichomes secrete powerful stinging compounds; deter even large animals

Leaf Internal Structure

- Ground tissues interior to leaf epidermis—**mesophyll**
- Along upper surface of most leaves—**palisade parenchyma** (aka **palisade mesophyll**)
  - Main psn tissue
- Lower portion of leaf—**spongy mesophyll**
  - Open, loose parenchyma, permits CO_2 and O_2 diffuse rapidly from stomata into all parts of leaf interior

- **Vascular bundles**, continuous with bundles in stem, located between palisade parenchyma and spongy mesophyll
- **Bundle sheath:** sheath of fibers may be around vascular bundle; flexible support
- Larger veins often have supportive fibers above and/or below: **bundle sheath extension**; ↑ support

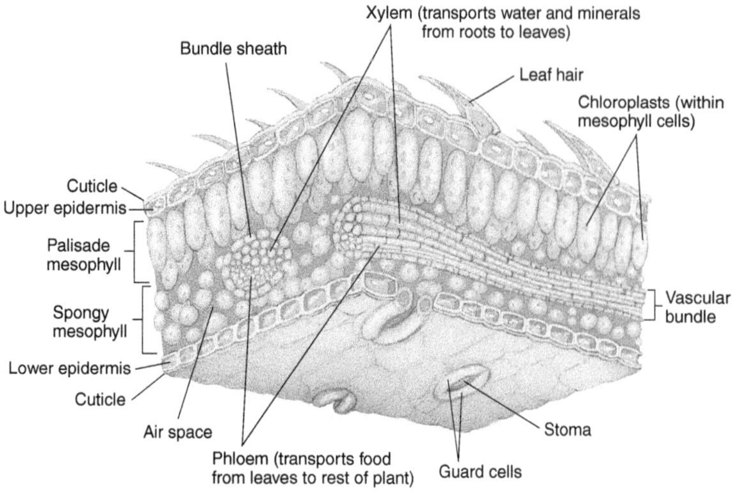

- **How Stomata Open/Close**
  - **Stoma:** (opening) surrounded by **guard cells**
  - Plenty of H_2O = guard cells open stoma (↑ transpiration and gas exchange)
  - Less H_2O = guard cells close stoma (↓ transpiration and gas exchange)
  - Same in leaf and stem

- Some of best support for Cohension-Tension theory: H_2O moves up stem even without roots!
- Many perennial plants, leaves initiate in summer or autumn
  - Go dormant as resting terminal or axillary bud until growing season begins
- When bud opens, young leaves rapidly expand and mature

- **Succulent Leaves**
  - desert habitats
  - thick and fleshy
    - = ↓ surface to volume ratio, H_2O conservation
  - Mesophyll: very few air spaces
    - = ↓ stomatal H_2O loss
  - Mesophyll more transparent; psn may occur deeper in leaf

- **Sclerophyllous Foliage/ Leaves**
  - aka "sclerophylls"
  - 2° walls expensive = try to ↓ amount sclerenchyma to save energy
  - for perennials, sclerophylls feasible
    - durable and highly protected
    - Examples: Holly, Barberry, etc.

- Conifer Leaves
  - sclerophylls
  - thick cuticle
  - epidermis and hypodermis (second epidermal layer under epidermis) cells have thick walls = ↓ H_2O loss
- Usually contain unpalatable chemicals
- Shapes: needlelike, flat, and scalelike
- Mostly perennial, remain on stem for many years; "Evergreens"

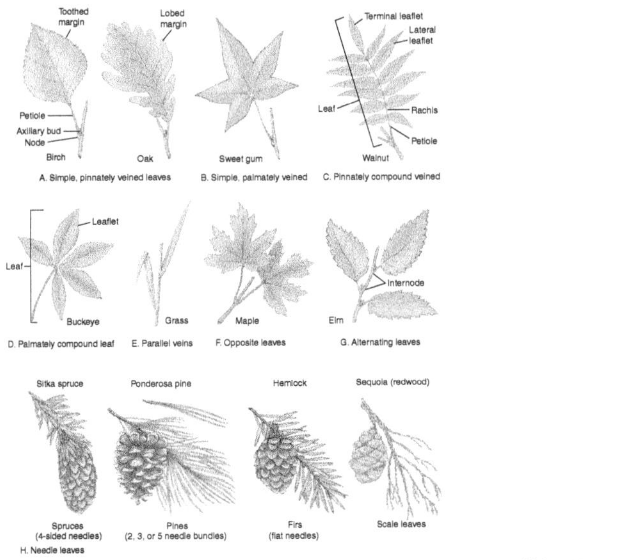

- **Bud scales**—tight layer around stem tip
  - Protect dormant buds in winter
  - Short or absent petioles
  - Small, very waxy, and tough
  - May have thin, corky bark layer

Spines, Thorns, and Prickles

- **Spine:** modified leaf, allows for defense; mostly collection of fibers
- ***Thorns:** a modified stem arising from axil of a leaf
- ***Prickle:** sharp outgrowth of epidermis or cortex

*Not really a leaf, but included for clarity

- Tendrils
  - Modified leaves whose cells can
    - sense contact with objects
    - coil around objects
    - support thallus
  - Grow continuously, do NOT psn

- **Insect Traps**
  - Habitats poor in usable N
  - Leaves highly modified to acquire animals for N
- Trap leaves classified as:
  - Active traps, move during capture
  - Passive traps, no movement

Preparing for Drought/Winter

- **Abscission zone** forms at leaf base = cells involved in cutting off leaf as becomes useless during Fall/Winter
  - store expensive elements (Mg, N, etc.) in roots, drop leaves to save H_2O and energy
  - leaves "change color" = as accessory pigments broken down = reveal chlorophyll

- **Transpiration:** evaporative H_2O loss from shoot
  - serious problem if soil becomes very dry
  - important in planetary H_2O cycle, weather!
- In most leaves, number of stomata much > in lower than upper epidermis; helps:
  - ↓ H_2O loss (sunlight heats upper leaf surface more)
  - prevent fungal spores infecting leaf
- **Leaf scar:** at petiole base, exposed cells filled with suberin
  - = protect thallus from H_2O loss and infection

Photosynthesis and Cellular Respiration

The Basics

- **Photoautotrophs** gather energy from light to assimilate small organic molecules
- Must build all of own molecules using just CO_2, H_2O, and various elements from soil
- Plants, algae, some protists, and bacteria
- **Heterotrophs** cannot do this; take in organic molecules (eat) and respire them for energy; also includes completely parasitic plants

Energy and Reducing Power

- Energy carriers important to both psn and respiration
  - Light captured by plant pigments
    - Pigments convert light energy into chemical energy (adenosine triphosphate [ATP])
    - ATP phosphate bonds store energy
  - Each ATP molecule recycled/reused repeatedly, 1000 times per second

ATP: The Basics

- **ATP**: molecule in which cells utilize for energy transfer/use
- Glucose → ATP or energy not usable; use cellular respiration to do this
- Psn allows plant to make carbohydrate from CO_2 → carbohydrate → cellular respiration

- Reducing power
  - CO_2 is **oxidized** form of carbon
    - (oxidized = not carrying as many e^- as it could)
  - e^-(s) added to atom = **reduced**
  - Every oxidation occurs with a reduction
    - = full reaction is "oxidation–reduction reaction" or "redox reaction"
  - Organisms need both energy and **reducing power**: ability to attach electrons onto compounds

- Optimum solution for moving and handling reducing power (electrons) = use small molecules that are semistable and mobile:
  - Nicotinamide adenine dinucleotide (NAD^+)
  - Nicotinamide adenine dinucleotide phosphate ($NADP^+$)
- Both NAD^+ and $NADP^+$ are **oxidizing agents** - they oxidize molecule they react with
- Process forms NADH and NADPH, two strong **reducing agents**
- Tendency to accept or donate electrons (e^-) varies greatly = a molecule's **redox potential**

- Other e^- carriers
  - **Cytochromes** = intrinsic membrane proteins
    - contain a cofactor, heme, iron atom
    - carry e^- only between sites extremely close together within a membrane
  - **Plastoquinones**, such as cytochromes, transport e^- over short distances within membrane
  - **Plastocyanin**, small protein that carries electrons on copper atom
    - loosely associated with chloroplast membranes

Psn: The Light-Dependent Reactions

- Nature of light
  - Light = one small segment of **electromagnetic radiation spectrum**, which includes radiation of different wavelengths and amounts of energy

- **Pigment** = molecule that absorbs specific wavelengths of light
- When light energy absorbed by pigment, an e^- **activated**/raised to orbital of higher energy level (goes from **ground state** to an **excited state**)
- Chlorophyll a, absorbs only some red + blue wavelengths of visible light, reflects green wavelengths

- **Absorption spectrum** shows which wavelengths most strongly absorbed by a pigment
  - **Absorption spectrum** of chlorophyll a - it absorbs red light and blue light very well, green poorly
  - **Action spectrum** shows which wavelengths most effective at powering a photochemical process

- **Accessory pigments** = molecules that absorb wavelengths <u>not</u> absorbed by chlorophyll *a*
  - most common accessory pigments in land plants: chlorophyll *b*, carotenoids
- absorbed energy passed to chlorophyll *a*
- transfer, called **resonance**, allows chlorophyll *b* to absorb wavelengths that chlorophyll *a* cannot

Psn: The Big Picture

- Light-dependent reactions: convert light energy into ATP + NADPH
- Light-independent reactions: use ATP + NADPH to convert CO_2 to glucose + regenerate ribulose-1,5-bisphosphate (RuBP)

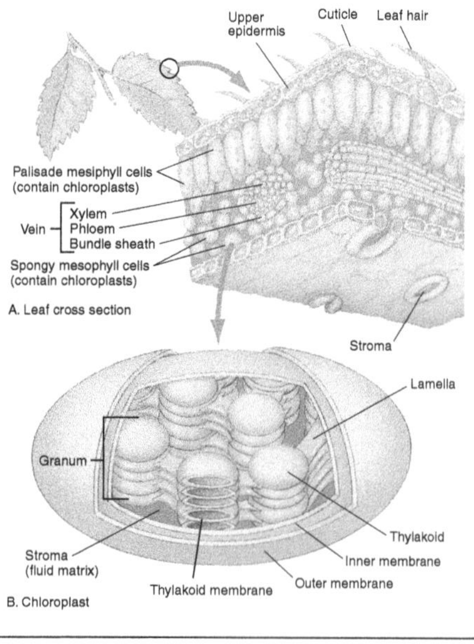

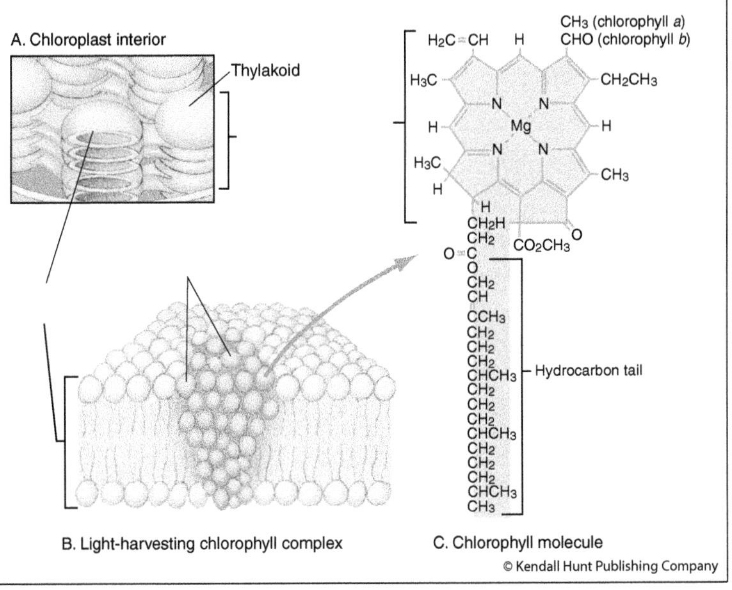

- Photosystem I (first identified)
  - All pigments and carriers work together, many millions in thylakoid membranes!
    - those with little chlorophyll *b*: **photosystem I** (first identified)
    - those in which chlorophyll *b* is present at levels almost equal to chlorophyll *a*: **photosystem II** (second identified)

- **Photosystem II** = mechanism that reduces P700, allowing it to work repeatedly
  - First, molecule of **plastocyanin** donates an e^- to chlorophyll a of photosystem I reaction center
  - It receives its new e^- from **cytochrome b6/f complex**, which gets an e^- from a molecule of plastoquinone

- Protons from light reactions build up in thylakoid lumen stream across membrane into stroma through special channels, complex sets of enzymes, **ATP synthetase**
  - ATP synthetase phosphorylates adenosine diphosphate (ADP) → ATP using power from proton flow

- When e^- flow smoothly from water to NADPH, the process is called **noncyclic e^- transport**
- Too little ATP produced this way relative to amount of NADPH produced
- After e^- reach ferredoxin in photosystem I, it can be transferred to plastoquinones of photosystem II instead of being used to make NADPH

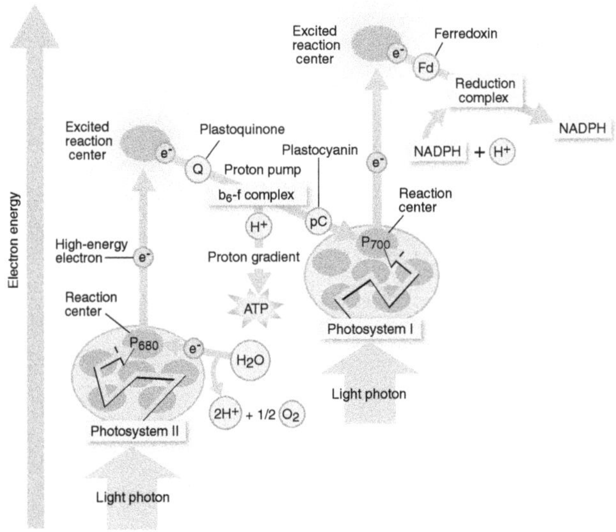

Cyclic e^- Transport

- Plastoquinones carry e^- along just as though they had gotten them from Q
- They use their energy to pump more protons into thylakoid lumen
- In this **cyclic e^- transport**, chloroplasts make extra ATP needed for stroma reactions

Light-Independent Reactions

- Conversion of CO_2 to carbohydrate occurs in stroma reactions; aka **Calvin-Benson cycle or C_3 cycle**
- First, **acceptor molecule RuBP** reacts with CO_2 molecule
- New molecule produced breaks into two identical molecules each containing three carbons: **3-phosphoglycerate** (PGA)
- Enzyme that carries out this reaction is **RuBP carboxylase (RUBISCO)**

- Second, ATP donates a high-energy phosphate group to 3-phosphoglycerate, converting it to **1,3-diphosphoglycerate**
- This then reduced by NADPH to **3-phosphoglyceraldehyde** (PGAL)
  - a phosphate comes off in this step
- The carbon is now both reduced and energized
- In short, PGA modified by enzymes using ATP + NADPH to make glucose and to regenerate RuBP

- As they operate, some molecules can be taken out of chloroplast and used by cell to build sugars, fats, amino acids, nucleic acids, and anything cell needs

Water

- Most plants keep stomata open during day, yet also lose H_2O
- Numerous structural modifications help conserve H_2O
- Metabolic adaptations to psn also exist:
  - C_4 metabolism
  - Crassulacean acid metabolism (CAM)

- In C_4 psn, CO_2 absorbed, transported, and concentrated in cells = more efficient
- O_2 kept away from RuBP carboxylase
- Occurs in leaves with Kranz anatomy
  - Each vascular bundle has a prominent chlorenchymous sheath of cells where majority of psn occurs

- Photorespiration rate ↑ with temperature, so any selective advantage of C_4 metabolism depends on environment
- Under warm dry conditions, C_4 metabolism MUCH more efficient than C_3 metabolism
  - Much less H_2O lost during CO_2 absorption
  - Examples: crabgrass, sugar cane, etc.

Crassulacean Acid Metabolism

- **CAM**, second metabolic adaptation that improves H_2O conservation while permitting psn.

- CAM metabolism almost identical to that in C_4 plants EXCEPT:
  - Malate (or other acids produced from oxaloacetate) not transported
  - It accumulates, in effect storing CO_2
  - However, occurs only at night, when stomata open

- Stomata closed during hottest periods, open only at night when cool = ↓ transpiration
- Daytime, when stomata closed, malate or other acids break down, releasing CO_2 for C_3 metabolism
- CAM not particularly efficient
  - Total amount of CO_2 so small that it may be entirely used after just few hours of sunlight

- CAM selectively advantageous in hot, very dry climate where survival rather than rapid growth most important
- Under milder, moister conditions, CAM not selectively advantageous = better to be C_4
- All CAM plants have other metabolic and structural adaptations for water conservation, such as:
  - Succulence
  - Tough epidermis with thick cuticle and wax layer

Overview

- **Respiration** breaks complex carbon compounds into simpler molecules and simultaneously generates ATP which cell requires
- Carbon oxidized as e^- removed by $NAD^+ \longrightarrow$ converted to NADH in the process
- NADH oxidized to generate more ATP

- NADH carries e^- to an e^- transport chain, which deposits them onto O_2, thereby reducing it and attracting protons to form H_2O
- Energy conserved as high-energy phosphate bonding orbitals of ATP
- Intermediate compounds of respiration's many steps useful as starting materials for many anabolic pathways

Types of Respiration

- **Aerobic respiration** requires O_2 as terminal e^- acceptor
- Most organisms: **obligate aerobes** = must have O_2 for respiration
- O_2 not available: alternative electron acceptor must be used = **anaerobic respiration/ fermentation**
  - Some bacteria- **obligate anaerobes** and killed by O_2
- Fungi and certain animal and plant tissues in animals: **facultatively aerobic**
  - can switch between types if O_2 is present/absent

Anaerobic Respiration

- Glucose broken down during anaerobic respiration by **glycolysis/Embden-Meyerhoff pathway**
- Once pyruvate formed and no O_2 present, all energy possible has been removed
- From each molecule of glucose= net production of 2 ATPs

- In plants and fungi, pyruvate first converted to **acetaldehyde**, then NADH reacts with that, forming **ethanol** (toxic if accumulated)
- Although method is far from ideal, it does allow certain organisms to survive in particular environments

Aerobic Respiration

- O_2 usually abundant and inexpensive
- Acts as terminal e^- acceptor
- Aerobic respiration consists of three parts:
  1. Glycolysis
  2. Citric acid cycle
  3. Oxidative phosphorylation in e^- transport system/chain (ETS)

1. Glycolysis to pyruvate occurs just as aerobic respiration
2. Pyruvate now be used in various metabolic pathways, but often broken down to generate ATP in **citric acid cycle/Krebs cycle/tricarboxylic acid cycle** to generate ATP

Citric Acid Cycle

- First, pyruvate moved into mitochondrial matrix, where oxidized and decarboxylated to form **acetyl**, which joins with **coenzyme A** to produce **acetyl CoA**
- Second, acetyl CoA enters citric acid cycle

Development and Morphogenesis

Concepts

- Differentiation requires that each cell
  - identify its position relative to other cells
  - inform other cells of developmental and morphogenic changes it is undergoing
- A **pattern establishment mechanism** in some cells informs each cell of its location relative to other cells

Environmental Complexity

- Most organisms exist in a heterogeneous environment – always changing
- It is selectively advantageous for organisms to sense differences and changes ⟶ respond to them

- Most plants are so large that their thallus exists in several different microenvironments.
- Intercommunication and coordination also needed at all levels of the thallus: cells ⟶ entire organism

Basic Mechanism for Communication

1. Plants must **perceive** environment cues or nucleus must receive chemical signals when conditions in surrounding cytoplasm change
2. Information must be **transduced** = changed to form that can be either acted upon or transported
3. There must be a **response** (e.g., dormancy, production of flowers, change in the type of leaf production, etc.)

Environmental Complexity

- Sensory systems and response mechanisms of plants tend to be simple
- Their signal transport is usually slow, involving movement through cortex parenchyma or phloem

Light

- **Direction**/gradient of light allows plant to orient its leaves toward source
- **Duration** of light (day) provides information about time of year

Gravity

- Selectively advantageous for plants orient selves/structures with respect to direction of gravity
  - Roots need to grow downward to access water and minerals
  - Shoots must grow upward to encounter better conditions for psn, pollination, and seed distribution

Touch

- Plant parts frequently grow against objects and respond to contact
- can use them for support or change direction to avoid them

Temperature

- Cold temperatures required for normal flowering of biennial and many perennials
- Low temperatures required to induce deep dormancy in temperate trees and shrubs

Water

- Presence of H_2O probably does not act like a signal in way other factors do
- Roots never actually grow toward water
- Water scarcity triggers plant responses that help plant cope with water stress

Responding to Environmental Stimuli

- **Tropic Responses** = growth responses oriented with regard to the stimulus
- **Positive tropism** = growth toward the stimulus
- **Negative tropism** = growth away from the stimulus
- **Plagiotropism** = growth at an angle

- **Phototropism** = growth in response to light
- **Thigmotropism** = growth in response to touch, exhibited by tendrils
- Pollen tubes of flowering plants may display positive **chemotropism**
  - Grow along style to ovary, following gradient of chemical(s) released from ovule

- Changing direction involves **differential growth**
  - One side of tendril must grow more than other
  - This can occur by:
    1. Growth on contact side slowing
    2. Growth on opposite side accelerating
    3. Both

Morphogenic Responses

- **Morphogenic response** causes change in metabolism of tissue or even entire plant
- **Presentation time** = length of time stimulus must be present for perceptive cells to react and complete transduction
- **Threshold** = level of stimulus that must be present during presentation time to cause perception and transduction

Chemical Messengers

- **Hormones** = organic chemicals produced in one part of thallus and transported to other parts, where initiate a response
- Most plant communication done via hormones

- Hormones act at very low concentrations
- Appear to be released into general circulation and not carried specifically to target
- Receptor cell contains almost all of information necessary for proper response
    - Hormones serve only to activate response

Auxin activation and inhibition of shoots

- As shoot apical meristems grow and initiate the new cells, also produces auxin indole-3-acetic acid (IAA)
- Auxin flows downward and surrounds all stem cells, causing the following three unique responses:
  - Cell elongation
  - Apical dominance
  - Differentiation of vascular tissues

Cell Elongation

- Auxin triggers cell elongation in cells of young internodes just below apical meristem
- At lower internodes, auxin not cause any response in fully grown, mature cells
  - They apparently lack proper auxin receptors

Apical Dominance

- Apically produced auxin induces dormancy in axillary buds
  - Each shoot tip gets only one active apical meristem = **apical dominance**

Apical Dominance

- As terminal shoot apical meristem grows away, auxin concentration around axillary bud gradually ↓ until drops below threshold; growth inhibition not maintained
- Axillary bud becomes active and grows out as branch or flower

Differentiation of Vascular Tissues

- Third site of auxin response in shoot tips: vascular cambium = cell division and morphogenesis
- Springtime, buds become active, produce auxin ⟶ stimulates cambial cells begin mitosis and cytokinesis; causes new daughter cells to differentiate into xylem

Interactions of Hormones in Shoots

- In some species, apical dominance is established by interplay of two to three hormones
- Active roots synthesize cytokinins, transported to shoot, and stimulate axillary buds
- Ratio of cytokinin and auxin determines whether buds are active or remain dormant

- A second hormonal interaction in vascular cambium:
  - Auxin alone activates cambium and elicits differentiation of xylem
  - Gibberellin also present in healthy stem and causes some of new cells differentiate as phloem

Hormones as Signals of Environmental Factors

- Leaf Abscission:
  - Hormones communicate to various plant parts when one part has encountered an environmental change
  - If a leaf damaged by animal feeding or H_2O stress, auxin production ↓ to such a low level that abscission zone does not stay quiescent
  - Autumn conditions seem to stimulate production of ethylene
  - Ethylene ↓ auxin production and transport in time for abscission before winter

- Fruits prevented from abscising prematurely by presence and export of sufficient amounts of auxin through pedicel
- Auxin causes developing fruits to release ethylene → other aspects of ripening
- At maturity, high concentration of ethylene stimulates pedicel abscission zone = it overrides presence of auxin → fruit drops/abscises

Tropisms

- Blue = most effective wavelength for phototropism
- Pigment whose absorption spectrum matches that is small protein: phototropin
- Site of perception of blue light:
  - stem tip in some species
  - **coleoptile** (outermost protective leaf at stem tip) for others
- When exposed to light, darker stem side receives extra auxin; grows more rapidly, and bends toward light

- In positive gravitropism in roots, root cap is organ of perception
- After lower side of root detected, growth inhibitor transported to lower side of root cap and then into root, slowing growth on lower side

Photoperiodic Induction to Flower

- In some annuals, size only important factor in induction to flower
- In many species, **photoperiod** (day length) triggers transition to flowering condition
- **Short-day plants** bloom when days short (Spring or Fall)
- **Long-day plants** bloom when days long, (Summer)
- **Day-neutral plants** not respond to day length

DNA and RNA

Concepts

- Cells differ in shape, largely due to differences in their gene expression
- Each cell contains same genetic code
- Development follows underlying process of **differential activation of genes**

- **Nucleic Acids:** organic molecules composed of monomers of nucleotides; function as information carriers and energy carriers
- **Nucleotide:** monomer of nucleic acids; composed of a sugar (two types), phosphate and base (five types)

DNA Molecular Structure

- Deoxyribonucleic acid (DNA)—deoxyribonucleic acid – a long threadlike molecule with uniform diameter, but varied length
  - 46 DNA molecules in the nucleus of most human cells
    - total length of 2 m
    - average DNA molecule 2 in. long
- DNA and other nucleic acids are polymers of nucleotides
- Each nucleotide consists of
  - one sugar – deoxyribose
  - one phosphate group
  - one nitrogenous base
    - Either pyrimidine (single carbon-nitrogen ring) or purine (double ring)

- **Nucleotide:** monomer of nucleic acids; composed of a sugar (two types), phosphate, and base (five types)
- Two types:
  1. DNA: molecule of inheritance, double-stranded helix, deoxyribose sugar, nucleotides: Adenine, Thymine, Guanine, Cytosine nucleotides
     - A bonds to T, C to G, via hydrogen bonds

Nitrogenous Bases of DNA

- Purines - double ring
  - Adenine (A)
  - Guanine (G)
- Pyrimidines - single ring
  - Cytosine (C)
  - Thymine (T)
- DNA bases - ATCG

Complementary Base Pairing

- Nitrogenous bases united by hydrogen bonds
  - a purine on one backbone with a pyrimidine on the other
  - A – T two hydrogen bonds
  - C – G three hydrogen bonds
- DNA base pairing
  - A – T
  - C – G
- Law of Complementary Base Pairing
  - one strand determines base sequence of other

Discovery of the Double Helix

- By 1900: components of DNA were known
  - sugar, phosphate, and bases
- By 1953: x-ray diffraction determined geometry of DNA molecule by Rosalind Franklin
- Nobel Prize awarded in 1962 to three researchers: Watson, Crick, and Wilkins, but not to Rosalind Franklin, who died of cancer at age 37, after discovering thex-ray data that provided the answers to the double helix.

DNA Function

- Genes—genetic instructions for synthesis of proteins
- Gene—segment of DNA that codes for a specific protein
- Genome—all the genes of one person
  - humans have estimated 25,000 – 35,000 genes
    - 2% of total DNA
    - other 98% is noncoding DNA
      - plays role in chromosome structure
      - regulation of gene activity
      - no function at all – "junk" DNA

What Is a Gene?

- Previous definition–gene–a segment of DNA that carries the code for a particular protein???
  - Body has millions of proteins but only 35,000 genes?
  - Small percent of genes produce only RNA molecules
  - Some segments of DNA belong to two different genes
- Current definition–gene–an information-containing segment of DNA that codes for the production of a molecule of RNA that plays a role in synthesizing one or more proteins
- Amino acid sequence of a protein is determined by the nucleotide sequence in the DNA

Replication of DNA

- DNA replicated during S (Synthesis) phase of cell cycle
- Each strand of DNA acts as template for complementary strand
- **DNA polymerase** (DNA pol.): enzyme that adds DNA nucleotides to growing new DNA strand

- Each strand of DNA acts as template for complementary strand
- **DNA polymerase** (DNA pol): enzyme that adds DNA nucleotides to growing new DNA strand
- **Replication bubble**: unwinding of double helix so DNA pol can replicate strand
- New double helix = 1 new strand + 1 "old" strand = **semiconservative replication**

Mutations

- **Mutation:** any change, however large or small, in DNA base pair sequence
- Causes of Mutations:
  - **Mutagens:** agents that cause mutations
  - **Insertions and deletions:** + or – of bases
  - **Transposable elements (transposons):** DNA segment that causes insertion and deletion mutations, various sizes (genetic engineering tool)

Effects of Mutations

- Effect and significance of mutation depend on its nature, position, and size
- Mutation in coding region of gene (exon) may cause gene to code for protein with disrupted active site = nonfunctional
- Statistically, mutations almost always harmful
- Natural selection (NS) eliminates deleterious mutations and preserves beneficial ones

Somatic Mutations

- Mutations in cells that never lead to sex cells: **somatic mutations**
- Somatic mutations can be important if affect part of plant that gives rise to vegetative offshoot (strawberries, grasses, etc.) or gametogenic cells

DNA Repair Processes

- Eukaryotes have DNA repair mechanisms that recognize and correct mutations; Prokaryotes don't = more mutations!
- DNA repair mechanisms must not be too perfect/efficient or NO mutations, ever, and no differences for NS to act on

Storing Genetic Information

- Protecting genes
  - DNA must be safely stored or will become inaccurate, useless, or even harmful
1. DNA not participate directly in protein synthesis
   - **messenger RNA** (mRNA) carries info from DNA to site of protein synthesis

2. Most DNA stored in nucleus
3. **Histone proteins** hold most nuclear DNA in inactive form
   - Histones form aggregates
   - DNA/protein structure, **chromatin**, so dense that enzymes cannot penetrate it; inactive/inert

The Genetic Code

- Twenty AA used in synthesizing proteins
- DNA nucleotides read 3 at a time
  - Sixty-four possible triplets (**codons**) possible
- UAA, UAG, and UGA: **stop codons**
- AUG: **start codon**
- Remaining forty possibilities also code for AAs
  - Most AA have (+)1 codon = **degeneracy**; further protects DNA

Genetic Code Is Universal

- All organisms and genetic systems share genetic code
  - Mitochondrial DNA is exception
- Strongest evidence that life arose only once on Earth and all living organisms evolved from one ancestral organism

Structure of Genes

- Controlling molecule(s) can recognize gene, bind to it, and deactivate/activate it at proper time
- **Structural region:** codes for AA sequence
- **Promoter region:** regulates synthesis of mRNA from structural region

- Structural portion of genes contains two distinct types of regions:
  - **Introns:** nucleotide sequences <u>not</u> expressed (converted into AA)
  - **Exons:** nucleotide sequences whose codons eventually expressed as sequences of AAs in proteins

Transcription of Genes

- **Transcription:** RNA pol. binds at start signal and starts making RNA, continues until transcription stop codon

Protein synthesis

- **Ribosome:** large + small subunits of rRNA + proteins; in cytoplasm
- "Read" genetic message of mRNA, construct proteins using that information
- All cells have many ribosomes; so there are many copies of genes in DNA

- tRNA: folded up RNA strand that transports AA to growing polypeptide chain; in cytoplasm
- Are as many different tRNA types as codons
- Parts of tRNA:
  - **Anticodon:** complimentary to codon = AA attachment site

mRNA Translation

- **Translation:** synthesis of protein molecule by ribosomes based on mRNA codons
- **Initiation process:** involves start codon in A site; ribosomal subunits assemble, move start codon to P site, leaving A site free

- **Elongation:** tRNAs enter A site in ribosomal subunit, match complementary anticodons to mRNA codons
- If not match, leave; other tRNAs try A site
- If tRNA matches, ribosome subunits "slide over" moving new tRNA from A site to P site, allowing free A site to interact with other tRNAs
- Repeat
- Translation **terminates** when stop codon reached, ribosome subunits fall apart

UNIT THREE

UNIT III

Cell Division
Genetics
Evolution
Taxonomy

Molecular Evolution-Abiotic

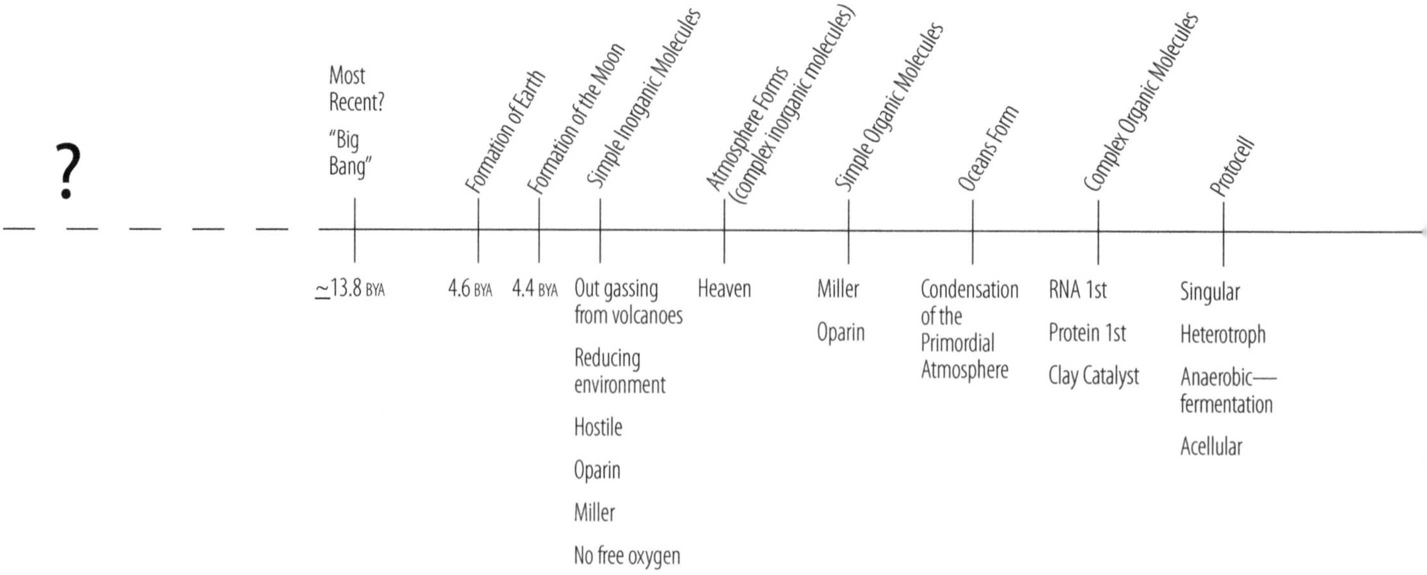

Cell Division

Maintaining the Chromosomal Number

- **Chromatin**—tangled mass of thin threads of DNA found in the nucleus (when the eukaryotic cell is not undergoing cell division)
  - Looks like angel hair pasta
- **Chromosomes**— chromatin coils, loops, and condenses to become these highly compacted structures (during cell division)
  - Looks like macaroni

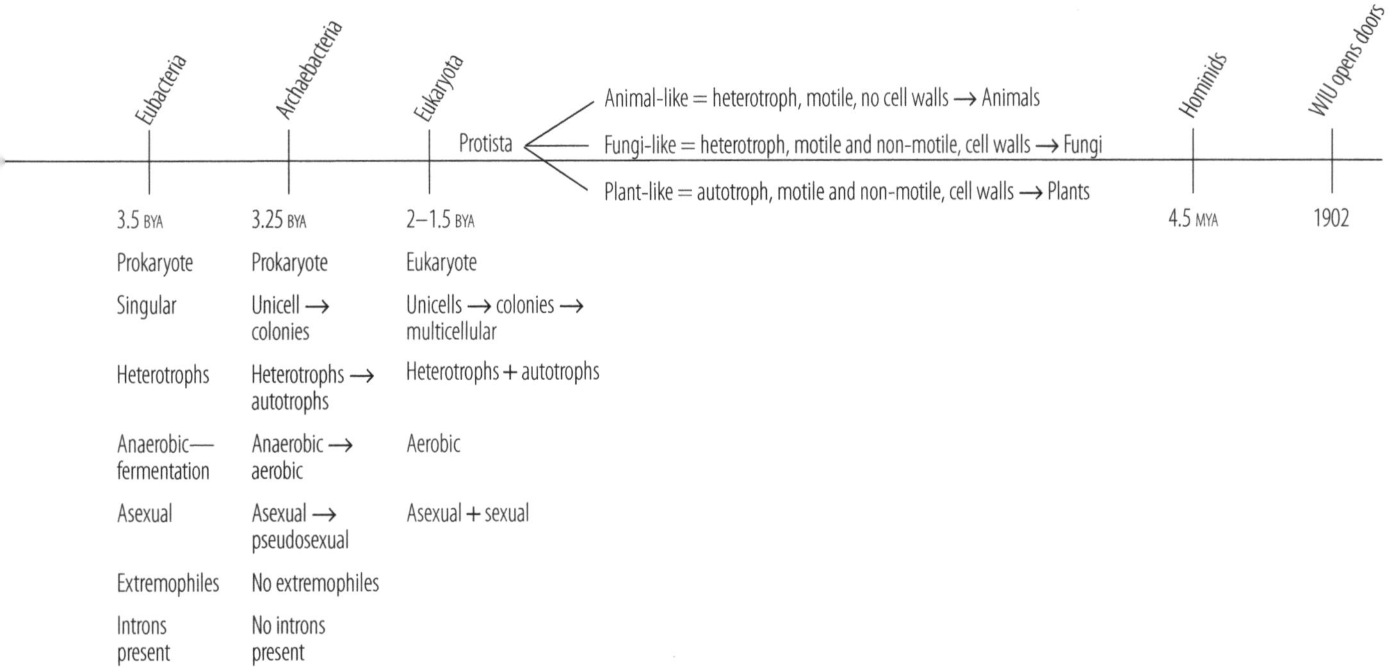

Cell Division

- **Cell Division**—involves nuclear division and cytokinesis
- **Cytokinesis**—division of the cytoplasm
- **Mitosis**—nuclear division in which the chromosome number remains constant
  - Occurs in growth and repair of the body

- Cell Cycle—the cell's life cycle that extends from one division to the next
- G_1 phase, the first gap phase
  - interval between cell division and DNA replication
  - accumulates materials needed to replicate DNA
- S phase, synthesis phase
  - duplicates centrioles
  - DNA replication occurs
- G_2 phase, second gap phase
  - interval between DNA replication andcell division
  - finishes centriole duplication
  - synthesizes enzymes that control cell division
  - repairs DNA replication errors
- M phase, mitotic phase
  - cell replicates its nucleus
  - pinches in two to form new daughter cells
- *Interphase*—collection of G1, S, and G2 phases
- *G_0 (G zero) phase* - cells that have left the cycle for a "rest"
  - muscle and nerve cells
- Cell cycle duration varies between cell types

Mitosis

- Cell division in all body cells except the eggs and sperm
- Functions of mitosis
  - development of the individual from one fertilized egg to some forty trillion cells
  - growth of all tissues and organs after birth
  - replacement of cells that die
  - repair of damaged tissues
- Four phases of mitosis
  - prophase, metaphase, anaphase, telophase

Mitosis: Prophase

- Chromosomes shorten and thicken coiling into compact rods
  - easier to distribute to daughter cells than chromatin
- Forty-six chromosomes
  - two chromatids per chromosome
  - one molecule of DNA in each chromatid
- Nuclear envelope disintegrates and releases chromosomes into the cytosol
- Centrioles sprout elongated microtubules – spindle fibers
  - push centrioles apart as they grow
  - pair of centrioles lie at each pole of the cell
- Some spindle fibers grow toward chromosomes and attach to the kinetochore on each side of the centromere
- Spindle fibers then tug the chromosomes back and forth until they line up along the midline of the cell

Mitosis: Metaphase

- Chromosomes are aligned on cell equator
  - oscillating slightly and awaiting signal that stimulates each of them to split
- Mitotic spindle—lemon-shaped array of spindle fibers
  - long spindle fibers (microtubules) attach to chromosomes
  - shorter microtubules (aster) anchor centrioles to plasma membrane at each end of cell

Mitosis: Anaphase

- Activation of an enzyme that cleaves two sister chromatids apart at centromere
- Daughter chromosomes migrate toward each pole of the cell with centromere leading the way
  - motor proteins in kinetochore crawling along the spindle fiber as the fiber itself is "chewed up" and disassembled at the chromosomal end
  - daughter cells of mitosis are genetically identical

Mitosis: Telophase

- Chromatids cluster on each side of the cell
- Rough endoplasmic reticulum produces new nuclear envelope around each cluster
- Chromatids begin to uncoil and form chromatin
- Mitotic spindle breaks up and vanishes
- Each nucleus forms nucleoli
  - indicating it has already begun making RNA and preparing for protein synthesis

Cytokinesis

- Cytokinesis—the division of cytoplasm into two cells
  - telophase is the end of nuclear division but overlaps cytokinesis
  - early traces of cytokinesis visible in anaphase
- Achieved by motor protein myosin pulling on microfilaments of actin in the terminal web of cytoskeleton
- Creates the cleavage furrow around the cell equator
- Cell eventually pinches in two

Timing of Cell Division

Cells divide when:

- they have enough cytoplasm for two daughter cells
- they have replicated their DNA
- adequate supply of nutrients
- are stimulated by growth factor
  - chemical signals secreted by blood platelets, kidney cells, and other sources
- neighboring cells die, opening up space in a tissue to be occupied by new cells

Cells stop dividing when:

- snugly contact neighboring cells
- when nutrients or growth factors are withdrawn
- contact inhibition—the cessation of cell division in response to contact with other cells

Foundation of Genetics

- **Gregor Mendel**—Austrian monk, developed laws of inheritance in 1860
- **Gene**—Unit of heredity that codes for a protein and is passed on to the offspring
- **Alleles**—Alternative forms of a gene having the same position on a pair of homologous chromosomes and affecting the same trait
- **Loci**—location of the gene on the chromosome

Mendel's Law of Segregation

- Each individual has two factors for each trait
- The factors segregate (separate) during the formation of the gametes
- Each gamete contains only one factor from each pair of factors
- Fertilization give each new individual two factors for each trait

Independent Assortment

- Each pair of factors segregates (assorts) independently of the other pair
- All possible combinations of factors can occur in the gametes

  *Dependent on the random arrangement of the homologous pairs at the metaphase plate of the spindle

Mendelian Cross

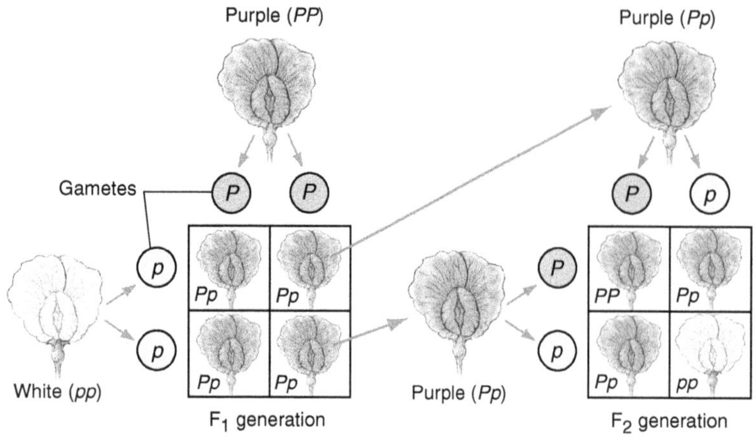

Gamete Formation

- The gametes have only one allele for each trait in accordance with Mendel's – Law of Segregation
- The member of each allelic pair separate during meiosis
- Gamete (haploid), therefore, contains exactly half the number of alleles as the parents cells (diploid)

Gamete Formation
Practice Problems

- How to determine the possible gametes that could be produce from an individual of known genotype...
- Example:
  RrWWTtbb = Parental Genotype
  (RWTb) (rWTb) (RWtb) (rWtb) = Possible gametes
- Example:
  RRWwttBB = Parental Genotype
  (RWtB) (RwtB) = Possible Gametes

Single Trait Inheritance

- Mendel suggests using letter to represent factors now called alleles
- **Dominant Trait**—indicated by the use of a representative capital letter
- **Recessive Trait**—indicated by the use of the appropriate lower case letter

Genotype

- Two letters used to represent the genes of the individual
- **Homozygous Recessive**—the two members of the allelic pair are the same and both recessive (ww)
- **Homozygous Dominant**—the two members of the allelic pair are the same and both dominant (WW)
- **Heterozygous**—the two members of the allelic pair are different (Ww)

Phenotype

- The physical characteristics of the individual—what the individual looks like as well as microscopic and metabolic characteristics
- **Example**:

Phenotype:	Genotype:
Widow's Peak	WW
Widow's Peak	Ww
Straight Hairline	ww

One Trait Crosses
Monohybrid

- Parental Generation (P_1)—Ww × Ww
- Gametes (W) (w) (W) (w)
- Punnett Square

	W	w
W	WW	Ww
w	Ww	ww

- First Filial Generation (F_1)
- Genotype: ¼ WW, ½ Ww, ¼ ww
- Phenotype: ¾ widow's Peak, ¼ Straight Hairline
- **Dominant Allele**—masks the expression of the recessive allele
- **Recessive Allele**—is expressed only when no dominant allele is present

Monohybrid Cross

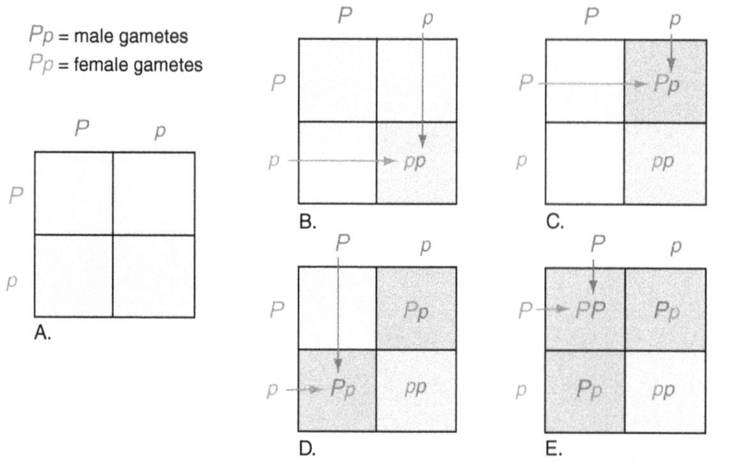

Probability

P_1 = Ww × Ww

- **Gametes:**

  The chance of (W) = ½

  The chance of (w) = ½

- **Genotype:**

  WW = ½ × ½ = ¼

  Ww = ½ × ½ = ¼

  wW = ½ × ½ = ¼

  ww = ½ × ½ = ¼

One Trait Crosses
Monohybrid Cross

- **Test Cross**—occurs when an individual with the dominant phenotype is crossed with and individual having the recessive phenotype

  * Determine the genotype of the parent with the dominant phenotype

Test Cross

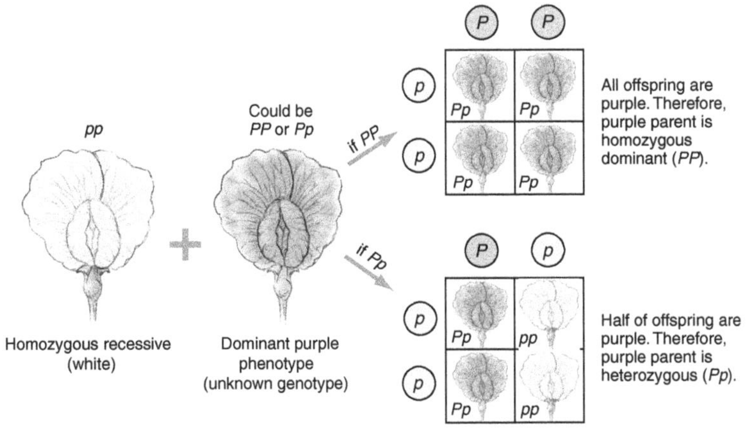

Two-Trait Crosses

- W = Widow's Peak w = Straight Hairline
- S = Straight Thumb s = Hitch-Hiker's Thumb
- P_1 = WWSS × wwss
- Gametes = (WS) (ws)
- F_1 = WwSs
- Gametes = (WS) (Ws) (wS) (ws)

Dihybrid Cross

Punnett Square

	WS	Ws	wS	ws
WS				
Ws				
wS				
ws				

Dihybrid Cross

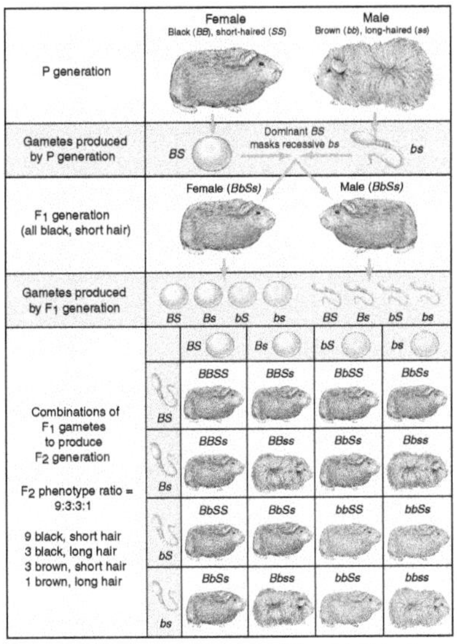

Two-Trait Cross and Probability

- Phenotype:
- ¾ Widow's Peak ¼ Straight Hairline

	W	w
W	WW	Ww
w	Ww	ww

Two-Trait Cross and Probability

- Phenotype:

  ¾ Straight Thumb ¼ Hitch-Hiker's Thumb

	S	s
S	SS	Ss
s	Ss	ss

Two-Trait Cross and Probability

- Therefore:
- The probability of the offspring having:
  - Widow's peak and Straight Thumb = $\frac{3}{4} \times \frac{3}{4} = \frac{9}{16}$
  - Widow's peak and Hitchhiker's Thumb = $\frac{3}{4} \times \frac{1}{4} = \frac{3}{16}$
  - Straight hairline and Straight Thumb = $\frac{1}{4} \times \frac{3}{4} = \frac{3}{16}$
  - Straight hairline and Hitchhiker's Thumb = $\frac{1}{4} \times \frac{1}{4} = \frac{1}{16}$
- Phenotypic ratio = 9:3:3:1

Polygenetic Inheritance

Epistasis

- Occurs when one trait is governed by two or more sets of alleles, possibly located on many different chromosomes
- Each dominant allele has a quantitative effect on the phenotype and these affects are additive
- The result is a continuous variation in phenotypes

Polygenetic Inheritance

Epistasis

- Skin Color
  - Unknown number of pairs of alleles (possibly as many as thirteen)
  - Few people have extreme phenotype
  - Most people have the phenotypes that lie in the middle

Polygenic Inheritance

- Genes at two or more loci, or even different chromosomes, contribute to a single phenotypic trait (skin and eye color, alcoholism, mental illness, cancer, and heart disease)

Multiple Alleles

- A trait is controlled by three alleles
- Each person has only two of the three possible alleles

Multiple Alleles

ABO Blood Type

- A = A antigen on the red blood cells
- B = B antigen on the red blood cells
- AB = A and B antigens on the red blood cells
- O = neither A nor B antigen on the red blood cells
- **Phenotype:** **Possible Genotype:**

Phenotype	Possible Genotype
A	I^AI^A, I^Ai
B	I^BI^B, I^Bi
AB	I^AI^B
O	ii

Rh Factor

Simple Mendelian Inheritance

- Inherited separately from A,B,AB, or O blood types
- Rh positive (Rh^+) is dominant over Rh negative (Rh^-)
- Rh^+ has a particular antigen on the red blood cells
- When an Rh^- woman reproduces with an Rh^+ man and the child in the womb is Rh^+, then the woman may begin to produce antibodies that will attack the red blood cells of the baby
- This can be prevented by giving the mother Rh immunoglobin injections midway during the pregnancy

Degree of Dominance

- Codominance–ABO blood types
- Incomplete Dominance–
  - HH = Curly Hair
  - HH′ = Wavy Hair
  - H′H′ = Straight Hair

Incomplete Dominance
Sickle Cell Disease

Genotype	Phenotype
$Hb^A\ Hb^A$	Normal
$Hb^A\ Hb^S$	Sickle Cell trait
$Hb^S\ Hb^S$	Sickle Cell positive

Incomplete Dominance
Sickle Cell Disease

- Sickle—shaped blood cells cannot pass along the narrow capillary passageways, clogging the vessels
- 60% of the population in Malaria—infected regions of Africa have the allele
- 10% of the African American population carry the allele

Incomplete Dominance
Sickle Cell Disease

- Among regions of Malaria infected —Africa
- Infants with sickle cell trait have a better chance of survival than the normal homozygote
- The protection afforded by the sickle cell trait keeps the allele prevalent in populations exposed to Malaria

Pleiotropy

- One gene produces multiple phenotypic effects
  - alkaptonuria—mutation on chromosome 3 that blocks the breakdown of tyrosine
  - sickle cell disease is another example of pleiotropy

Dominant and Recessive Alleles at the Population Level

- Common misconception that dominant alleles must be more common in the gene pool than recessive alleles
- Some recessive alleles, blood type O, are the most common
- Some dominant alleles, polydactyly and blood type AB, are rare in the population

Evolution
Fossil Evidence

- **Evolution**—the sum of all of the changes that have taken place in living things since the beginning of life
- **Fossils** —the remains and traces of past life
- Traces include:
    - Trails
    - Footprints
    - Burrows
    - Worm casts
    - Preserved droppings

Geologic Timescale
Relative Dating Method

- **Law of Superposition**—fossils are found in sedimentary rock which is laid down over long periods of time in layers (strata)
- The top layers are younger than those layers that are under them
- It can be estimated that the fossils found in these younger layers are also younger
- Does not give an age in years, but in relation to other fossils

Geologic Timescale
Absolute Dating

- Relies on radioactive dating techniques to assign an actual date to a fossil
- All radioactive isotopes have a specific Half-life
  - **Half-life**—the length of time taken for one half of the sample to change into another stable element
- Carbon-14 is the only radioactive isotope found in organic matter
- Very reliable up to 50,000 years

Geologic Timescale
Absolute Dating

- Certain radioactive isotopes can be used to date rocks directly, the age of the fossil in the rock can be inferred
- The ratio of Potassium 40 and Argon 40 trapped in rock is often used
- Used in rock less than 100,000,000 years old

Geologic Timescale

Absolute Dating

- Ratio of isotope Uranium 238 and Lead 207
- Has a very long half-life
- <u>Used in rocks older than 100,000,000 years</u>

Mass Extinctions

Theories

- **Extinction**—the death of every member of a species
- **Mass Extinctions**—times when a large percentage of the existing species become extinct in a relatively short time
  - There have been at least five mass extinctions
  - Occurred about every 26 million years

Mass Extinctions
Cretaceous Extinction

- The most well known
- The extinction of the dinosaurs
- Due to an asteroid that exploded, producing meteorites that fell to Earth
- A large meteorite striking the Earth could have produced a cloud of dust that blocked out the sun's warming light
- First plants died, then the largest herbivores, and finally the largest carnivores may have died in the cold

Tunguska Explosion July 30, 1908

- https://www.youtube.com/watch?v=EiXpp-i442s

Mass Extinctions

Cretaceous Extinction

- Some believe that less catastrophic events killed the dinosaur
- Possibly a virus
- Competition for space and resources caused overcrowding

Biogeographical Evidence

- **Biogeography**—the study of the distribution of plants and animals throughout the world
- Related forms evolve in one location and spread out into other regions
    - There are no rabbits in South America- they evolved some place else and had no means of getting there
    - There are no snakes found on certain islands for the same reason

Alfred Russel Wallace

- British naturalist, explorer, geographer, anthropologist, and biologist.
- Independently conceived the theory of evolution through natural selection
  - his paper on the subject was jointly published with some of Charles Darwin.
- Considered the 19th century's leading expert on the geographical distribution of animal species and is sometimes called the "Father of Biogeography."
- Unlike Darwin, he had no family wealth to fall back on

Intelligent Evolution

- Theory of common descent based on natural selection strictly bounded by the principle of utility
  - (i.e., the idea that no organ or attribute of an organism will be developed and retained unless it affords a survival advantage).
  - Where utility cannot be found in a known organ or attribute, some other cause—an intelligent cause—must be called upon.

Biogeographical Evidence
Continental Drift

- The position of the continents is not fixed
- They are constantly moving (Floating) on a sea of molten rock called Magma
- 225 million years ago—Pangea
- 135 million years ago—Laurasia and Gondwanaland
- 65 million years ago—present continents separate

Anatomical Evidence
Homologous Structures

- Structures that are similar were inherited from a common ancestor
  Example:
  - Forelimbs of mammals
- The wing of a bird and an insect are Analogous Structures (they are adapted for flying, but are structurally unrelated)

Animal Homology

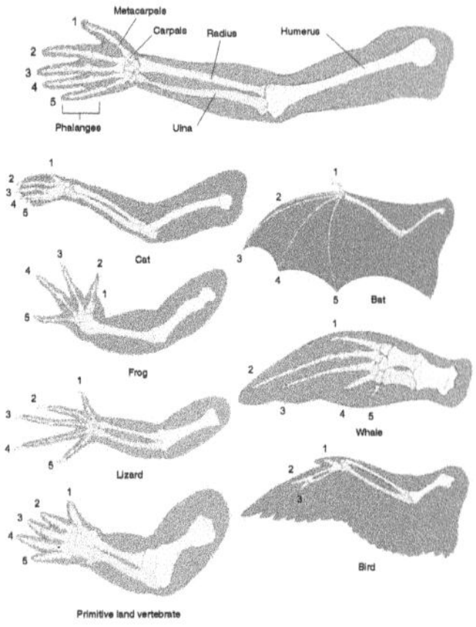

Anatomical Evidence

Homologous Structures

- Ernst Haeckel—(1834–1919)
- "Ontogeny recapitulates Phylogeny"
- The embryonic development of a species restates the many stages of evolution the species has gone through

Biochemical Evidence

- All organisms have certain biochemical molecules in common
- The degree of similarity between DNA and Amino Acid sequences is thought to indicate the degree of relatedness

Population Genetics

- **Population**—all members of a single species occupying a particular area at the same time
- **Gene Pool**—the various alleles of all the gene loci in all the members of a population
- **Hardy-Weinberg Law**—allele frequencies in a sexually reproducing population come to an equilibrium that is maintained generation after generation
  - No mutations
  - Random mating
  - No gene flow
  - No natural selection

Population Genetics
Hardy-Weinberg Law

- In real life, these conditions rarely if ever occur
- Allele frequencies do change
- The significance is that it tells us what factors cause evolution—those that violate the conditions listed

Agents of Evolutionary Change
Mutations

- Provide new alleles
- Underlie all other mechanisms that produce variation
- The raw material for evolutionary change
- The adaptive value of a mutation is entirely dependent on environmental conditions

Agents of Evolutionary Change

Genetic Drift

- Refers to the change in allele frequencies of a gene pool due to chance
- **Founder Effect**—occurs when only a few individuals found a colony and only a fraction of the total genetic diversity of the original pool is represented

  Example:
  - Pennsylvania Amish

Agents of Evolutionary Change

Gene Flow

- The movement of alleles between populations
- **Emigration**—the loss of genes in a gene pool by an individual leaving a population
- **Immigration**—the gaining of alleles in a gene pool by an individual joining a population

Agents of Evolutionary Change
Nonrandom Mating

- Occurs when individuals pair up according to their genotypes and phenotypes, not by chance

Agents of Evolutionary Change
Natural Selection

- Based, in part, on the work of Charles Darwin
- The process by which populations become adapted to their environment
- "Survival of the Fittest"
  - Herbert Spencer
  - fitness is measured by how reproductively successful its offspring are in the next generation

Agents of Evolutionary Change

Natural Selection

- **Variation**—the members of a population differ from one another
- **Inheritance**—many of these differences are heritable genetic differences
- **Differential Adaptedness**—some of these differences affect how well an organism is adapted to its environment
- **Differential Reproduction**—individuals that are better adapted are more likely to reproduce, and their offspring will make up a greater proportion of the next generation

Speciation

- A group of interbreeding subpopulations that share a gene pool and that are isolated reproductively from other populations

Premating

- Habitat (occupy different habitat)
- Temporal (reproduce at different time of year or day)
- Behavioral (courtship behaviors differ)
- Mechanical(genitalia unsuitable for one another)

Postmating

- Gamete Isolation (sperm cannot reach egg)
- Zygote Mortality (zygote does not survive)
- Hybrid Sterility (hybrid survives but is sterile)
- F2 Fitness (hybrid survives but has reduced fitness)

Modes of Evolutionary Change

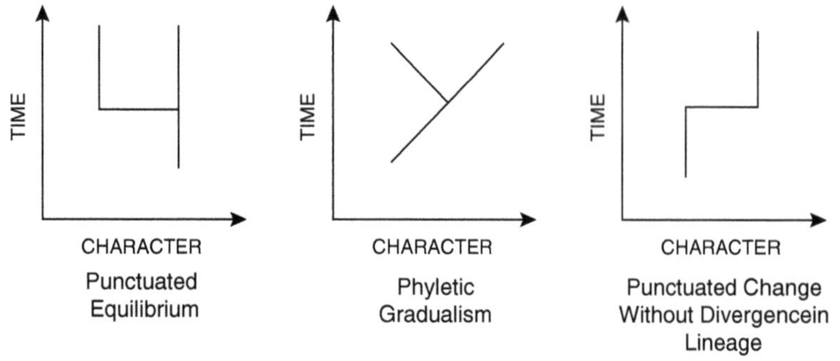

Naming and Organizing Plants and Microbes

Taxonomy

Binomial Nomenclature

- Carolus Linnaeus (1707–1778)
- Swedish biologist
- Invented the two-part naming system

Binomial System

- Made up of two names
  - Genus
  - Species

Binomial Rules

Genus

- Always first
- Always capitalized
- Must be *italicized* or <u>underlined</u>

Binomial Rules

Species

- Always second
- Always lower case
- Must be *italicized* or <u>underlined</u>
- **Example:** *Homo sapiens* or <u>Homo sapiens</u>

Basic Concepts

- **Taxon** (pl. = taxa): monophyletic group
- **Monophyly:** a group of organisms that has descended form a common ancestor
- **Phylogeny:** science of determining/studying hereditary/(evolutionary) relationships among taxa
  - Evolutionary history of taxon and its members
- **Systematics:** science and practice of organizing and classifying species
- **Classification(classn):** science of identifying, naming, and placing individuals/species/taxa in their proper groups

- Goals of systematics:
  - To understand evolutionary lineages
  - To maintain naming system (**nomenclature**) that reflects relationships accurately and is stable

- At the end of 19th century, taxonomists adopted goals:
  1. Developing natural system of classn, grouping closely related organisms/taxa
  2. Assigning plant names on the basis of phylogenetic relationships

- A taxon is **monophyletic** if:
  - all members related to each other by a common ancestor
  - all descendants of that common ancestor included
- Polyphyletic species may resemble each other only due to **convergent evolution**
- A taxon is **polyphyletic** if:
  - members have evolved from different ancestors

How Do We Know How Taxa Are Related?

- Analyze via phylogenetic studies
- Originally, only morphological data
- Then, anatomy, cellular, biochemical, and now, DNA data
- Major relationships worked out, now, mostly "detail work"

- Two reasons organisms can resemble each other:
  1. Descent from a common ancestor results in **synapomorphies (homologous features)**
  2. Convergent evolution results in **homoplasies (analogous features)**

- Studying lack of similarity difficult too:
  - a small genetic change → dramatic phenotypic changes
- Systematists attempt to examine as many features as possible, so that misleading data outweighed by informative data

Understanding Cladograms

- Cladogram: branched diagram of evolutionary relationships among taxa
- Each node represents ancestral group/divergence of one taxon → two taxa
- All taxa beyond "hash mark" possess that trait

- **Paraphyletic group:** not contain all descendants of most recent common ancestor (reptiles: not contain birds)
- **Apomorphy:** derived condition
- **Ancestral condition:** condition of ancestors
- **Synapomorphy:** shared, derived trait; used to construct cladograms
- **Parsimony:** principle of choosing simplest possible explanation for given situation

Cladograms and Taxonomic Categories

- Only taxonomic unit with an objective definition = species: individuals that can interbreed (works for animals only!)
- Many systematists have chosen to use the older names for genera, families, and orders unless proven to be nonmonophyletic (most parsimonious; avoid confusion; stability)

Other Types of Classification Systems

- Artificial systems of classification: Guides, Keys
  - Use several main easy to observe traits, chosen as basis of artificial classification (easy identification = main purpose)
  - "Artificial" because groups often NOT share close common ancestor (grouped by color, shape, etc.)

UNIT FOUR

UNIT IV

Angiosperm Diversity
Flowering Plants and Coevolution
Bryophytes
Gymnosperms

Angiosperm Diversity
and Reproduction

Image Courtesy of Tom Vogel

Flower Anatomy

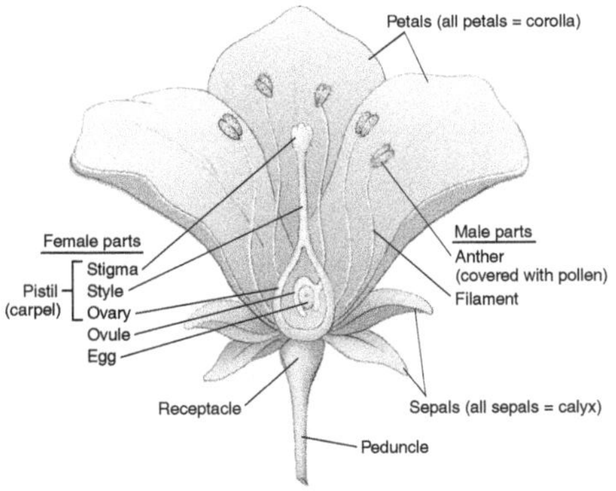

Flower Structure and Cross-Pollination

- **Cross-pollination** = fertilization by pollen from different individual
- **Self-pollination** = fertilization by pollen from same flower or another on same plant
- Results of each similar sexual versus asexual reproduction
- Mechanisms evolved to:
  - ↓probability of self-pollination
  - ↑chances of cross-pollination
  - ↑genetic diversity

- Stamen and style maturation times
  - Self-fertilization in flowers with both stamens and carpels prevented if anthers and stigmas mature at different times
  - Plants with many flowers—not open simultaneously, older flowers may be self-pollinated by newly opened flowers of same plant

Self-pollination may be inhibited by **compatibility barriers** = chemical reactions between pollen and stigma can prevent pollen growth

Angiosperm Reproduction

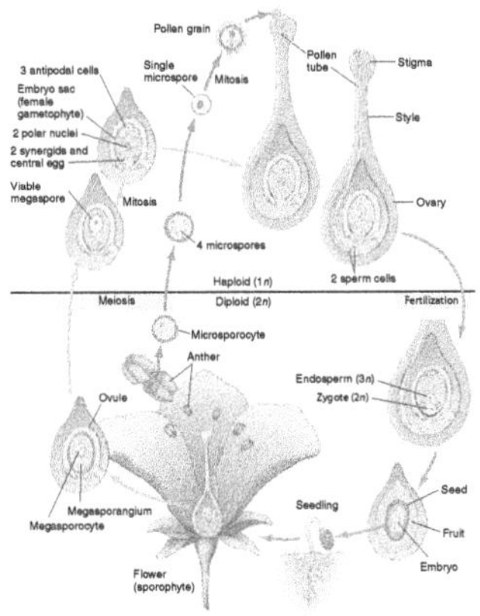

- Wind-pollinated flowers
  - usually no petals and reduced/absent sepals
  - huge number of pollen grains produced to ↑ chance of pollination
  - large, feathery stigmas ↑ area to catch pollen grains
  - 1000+ tiny flowers per plant
  - form dense populations to ↑ pollination
  - Examples: grasses, pines, oaks

Fruit Types and Seed Dispersal

- **Fruit**: a mature ovary
  - aids in seed dispersal in some way
- If animals to disperse seeds, part of fruit must be edible/ attractive, yet seed and embryo must be protected from consumption/digestion

- **Simple fruits**: develop from one to several fused carpels from single flower; fleshy or dry
  - Examples: berries, drupes, pomes, legume, capsule, samara
  - Dehiscent versus indehiscent
- **Complex fruits**: develop from multiple pistils or multiple flowers; mostly fleshy
  - Examples: aggregate fruits ("berries"), pineapples, Osage orange

- Wind-dispersed seeds are lightweight
    - wings/parachutes
- Fruits and seeds transported by water must be:
    - buoyant
    - resistant mildew and rot

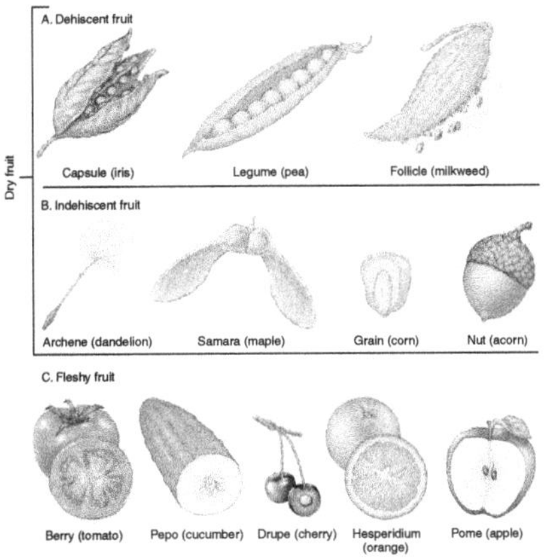

Germination Basics

- **Germination:** process of seed coat opening and allowing embryonic sporophyte to grow
  - radicle grows down; matures into root
  - epicotyl grows up; matures into shoot
- Each species requires special environmental requirements: H_2O, temperature, humidity, etc.
- **Seed coat**: tough protective covering of embryo and endosperm
  - tough, resistant, composed of sclerenchyma

Seed Anatomy

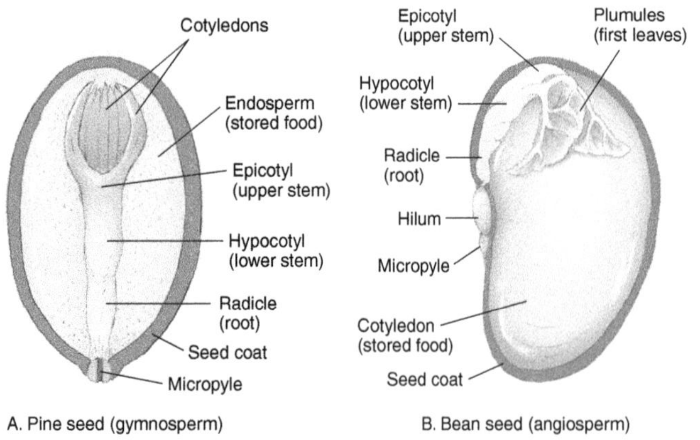

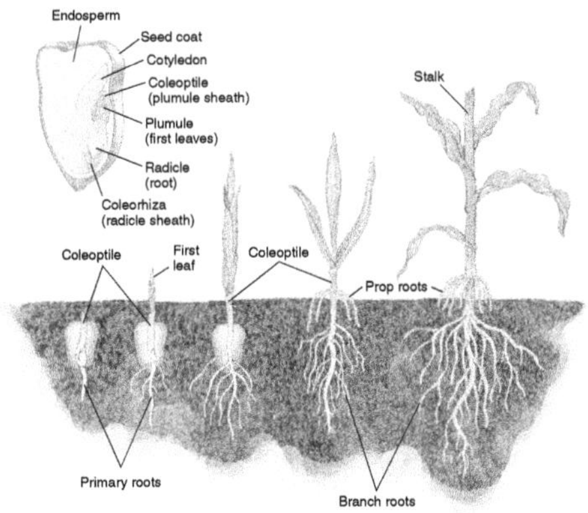

Flowering Plant and Animal Coevolution

The Basics

- **Coevolution:** reciprocal evolutionary changes two species cause each other because in such close association together
- Plants and animals have coevolved together, even before moved on to land
- Both are ancient groups

1. **Pollination:** transfer of pollen from one anther to one stigma via wind, animal, water, etc.
2. **Anti-predator mechanisms:** genetically controlled production of chemicals to deter or harm predator (a whole other book!) OR trade home for insect protection (*Acacia* and ants)
   - Acacia produces hollow thorns/homes for ants. Ants protect home from herbivores and provide fertilizer/poop

Wind/Water-Pollinated Flowers

- NO color, scent, etc., as lost nonessential traits to save energy
- Do produce 100k millions of pollen per plant, so have chance of reproducing

Pollination Coevolution

- Animal pollinators = successful pollination and seed production
- Avoids inbreeding (mating with relatives); causes ⇑ in genetic diversity
- Plant trades nectar and excess pollen (food) for transfer of pollen by pollinator

Image Courtesy of Tom Vogel

- Species may have (+)1 pollinator
- above 300k species animals are pollinators—birds, bats, insects, mammals, etc.
- Essential to global ecology—no plant reproduction = no animals

Some Details

- Flowers position, color, shape scent, bribe, etc., to attract their pollinator(s)
  - locate flowers where pollinator can detect (edge of tree, on top of stem, etc.)
  - **nectar**, pollen are high in energy (sugar) and amino acids – bribe pollinator
  - must be proper "color" for pollinator, so what humans "see" is not necessarily what pollinator "sees" in infrared or ultraviolet (UV)

Some Specifics

- Beetle: open, robust flowers, large and flat, white, LOTS of pollen, regular flowers
  Example: *Magnolia*
- Bee: landing platforms, UV/blue/purple/lavender/white (NOT red), with "bee guides", irregular flowers, nectar and pollen, sweet scent
  Examples: tomato, fruit trees, asters
- Flies: red/light color, stinky, nectar,
  Examples: *Titan*, *Raflesia*

- Moth: white, open at night, strong scent, nectar
  Examples: moonflower (*Ipomoea*), yucca
- Birds: (mostly tropical), a perch, red, usually little scent, open daytime, nectar

- Moth: white, open at night, strong scent, plenty of nectar
  Examples: moonflower, Yucca

Animal Dispersers

- Animals, very mobile, useful as dispersers of seeds: eat them, stick to fur/feathers/feet, and transported elsewhere
- Fruits change colors to indicate they are mature. . . which reduces the chance of the consumption of immature seeds
- Ripe fruits are intense colors, so animals easily see

Origin of Life
The Protocell

- A structure that had a lipid-protein membrane and carried on energy metabolism
- Acellular
- Abiotic
- Not assigned to one of the three domains

Prokaryotes

- Compose two kingdoms or domains (Archaea and Eubacteria)
- Are cellular
- Typically 1/1000 the volume of small eukaryotic cells
- Most are unicells , but some are colonies or filaments

Bacteria
Asexual Reproduction

- Binary Fission
  - Asexual process
  - Circular DNA replicates
  - A new cell wall grows between the DNA molecules
- Can occur every 20 minutes under optimal conditions
- Can also reproduce asexually by Budding and Fragmentation

Bacteria
Pseudo-Sexual Reproduction

- "False" Sex
- Not sexual reproduction
- Genetic material may transfer through three mechanisms

Bacteria
Pseudo-Sexual Reproduction

- **Transformation**—fragments of DNA from a damaged cell may be taken up by another cell
- **Transduction**—bacteriophages (virus) transfer bacterial DNA from one to the other
- **Conjugation**—genetic material may be transferred via pili
  - DNA is transferred between different mating strains
  - Exchange occurs between donor and recipient strains

Eukarya (Eukaryotes)

- Appear about 1.5–2.0 bya
- Include the following four Kingdoms
  - Fungi
  - Plants
  - Animals
  - Protists

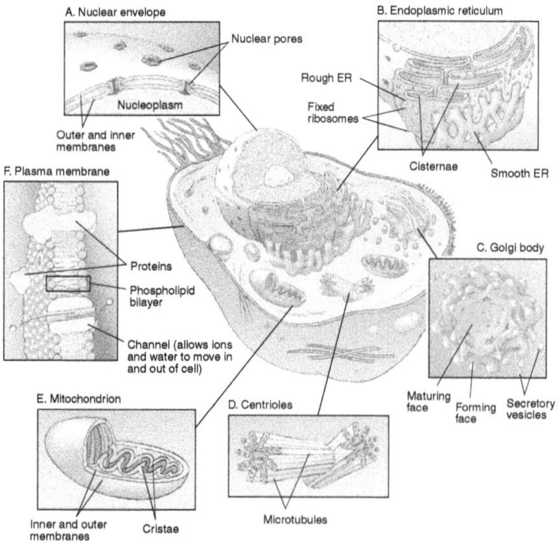

Multicellularity

- Some unicellular organisms began living in association with others
- Safety in numbers
- Take advantage of special abilities of some organisms
- Unicellular > Colonies > Multicellular

Kingdom Protista

General Ecology

- Vary in size, structure, mode of locomotion, and reproduction
- Most are mobile, and utilize cilia, flagella, or pseudopodia
- Polyphyletic group

General Ecology
Reproduction

- Range from asexual to sexual modes
- Most do not develop multicellular reproductive organs
- May reproduce through syngamy (union of gametes)

Evolution of Plants
Adaptation to Life on Land

- A cuticle prevents desiccation of plant tissue
- Stomata allow diffusion of carbon dioxide into the intercellular spaces of leaf tissue
- Multicellular gametangia have sterile cells as well as gametes
    - The fertilized eggs develop within the female gametangium

Bryophytes
Mosses, Liverworts, and Hornworts

- Seedless
- Disperse via haploid spores
- Diversified during the Silurian and Devonian periods
- Lack true vascular tissue
  - Limits them to moist environments
  - Restricts them to small size

Image Courtesy of Tom Vogel

Bryophytes

- Members of phylum Bryophyta are colonial plants
- Rhizoids anchor the plant to the soil
- Lack true leaves, roots and stems
- Some have conducting cells which function like vascular tissue

Mosses
Alternation of Generations

- **Gametophyte** is dominant
- Bears the gametangia at the apex of the plant
- Some have separate sexes
- Others bear archegonia and antheridia on the same plant

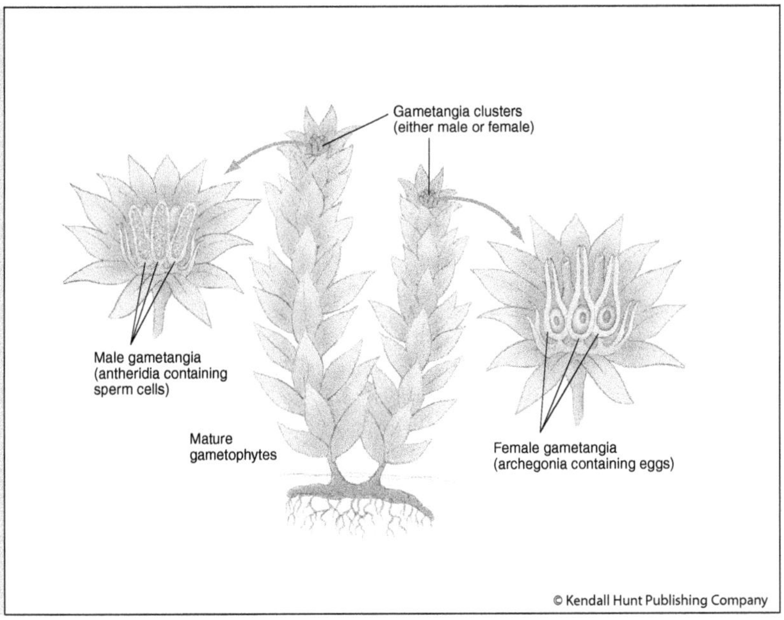

Mosses
Fertilization

- Flagellated sperm are transported to the archegonium by
  - splashing raindrops
  - providing the water for them to swim
- Sperm may also be transported by insects
- The archegonium secretes sucrose that attracts the insect

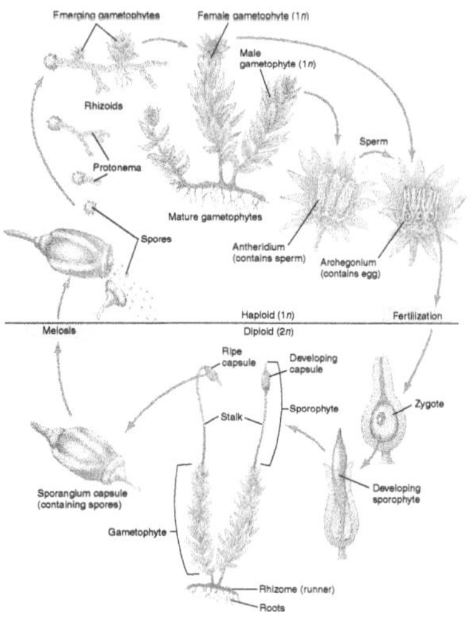

Mosses

- **Sphagnum**—the most commercially important moss
  - Cells of the gametophyte hold much water
  - Used as a soil conditioner and packing material

Hornworts

Inconspicuous Thalloid Plants

- Resemble liverworts but may not be closely related to them
- Have a single disc-shaped chloroplast in each cell (similar to many algae)
- Antheridia and archegonia are embedded in the thallus
- The sporophyte which develops after fertilization forms a "Horn"
- Meiosis occurs within sporangia
  - Spores are formed

Liverworts

- Phylum Hepaticophyta have a dominant gametophyte generation like mosses
- Body form is flattened
- Lobe-like thallus
- Rhizoids anchor the plant to the soil
- Sexual reproduction like that of mosses
- Asexual reproduction involves production of **Gemmate** that disperse via raindrops or small animals

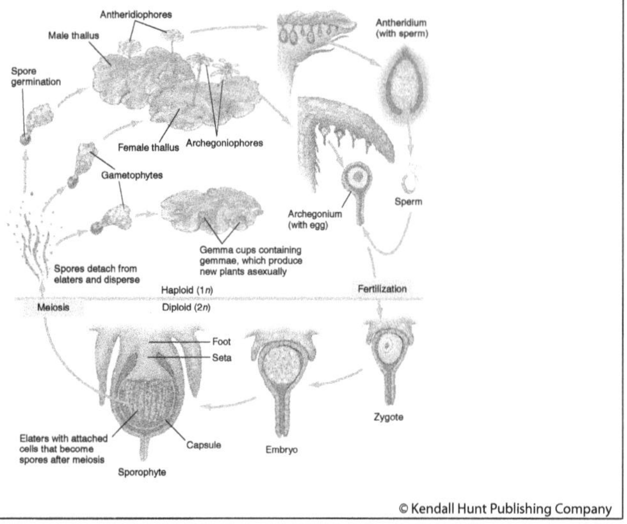

Mosses

Development

- The diploid zygote grows into a multicellular **Sporophyte** by mitosis
- The sporophyte grows on and is nutritionally dependent on the gametophyte
- The **Gametophyte** has a:
  - Foot
  - Seta(stalk)
  - Capsule -contains sporogenous (spore mother cells)

Ferns and Fern Allies

- 11,000 species of ferns
- Extant ferns are representative of what was once a much larger group
- Typically found in moist, tropical habitats
- A few are aquatic

Image Courtesy of Tom Vogel

Fern Allies
Whisk Ferns, Club Mosses, and Horsetails

- Have true vascular tissue (xylem and phloem) for conduction and support
- Most have true roots and leaves
- **Megaphylls** and **Microphylls** are two types of leaves

Fern Allies
Whisk Ferns, Club Mosses, and Horsetails

- Microphylls
  - Small
  - Have a single vascular strand
  - Probably evolved as extensions of stems
    - Club Mosses are the only extant group that possess microphylls

Fern Allies

Whisk Ferns, Club Mosses, and Horsetails

- Microphylls
  - Larger leaves
  - Have more than one vascular strand
  - Probably evolved from stem branchings
    - Ferns, horsetails, and seed plants have megaphylls

Ferns

Phylum Pterophyta

- Alternation of generations
- Dominant sporophyte generation
- Fern body consists of a **Rhizome** (an underground stem), roots, and leaves called Fronds
- **Fiddleheads**—newly emerging fronds
- All sporophyte structures contain vascular tissue

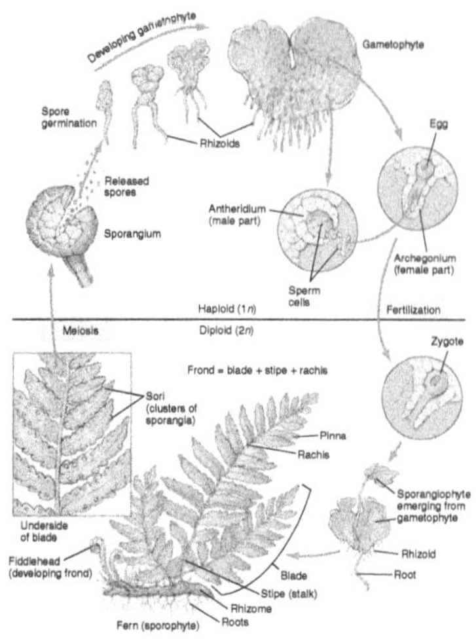

Ferns

Alternation of Generations

- Spore Production
  - Occurs on the underside of fronds in sporangia
  - Sporangia are found in clusters called **Sori**
  - Spores germinate and grow by mitosis into a gametophyte called the **Prothallus**

Ferns
Alternation of Generations

- Prothallus
  - Typically "Heart shaped"
  - Lacks vascular tissue
  - Anchored by Rhizoids
  - Bears antheridia and archegonia on the ventral surface
  - Water is still required for the transportation of the flagellated sperm to the archegonium
  - The zygote grows into the free-living sporophyte plant

Whisk Ferns
Phylum Psilotophyta

- The simplest vascular plants
- Lack true roots and leaves
- Have vascular stems that show dichotomous branching (primitive condition)
- Most species are extinct
- Extant species are limited to warmer climates

Whisk Ferns

Alternation of Generations

- Sporangia are born on the stems and produce spores
- Spores germinate to form underground, haploid prothalli
- Prothalli have a symbiotic relationship with mycorrhizal fungi

Horsetails

Phylum Sphenophyta

- Have hollow, jointed stems
- Were once among the dominant plants and grew as tall as modern trees
- Contributed to the coal deposits
- **Strobili**—the reproductive structures that bear sporangia

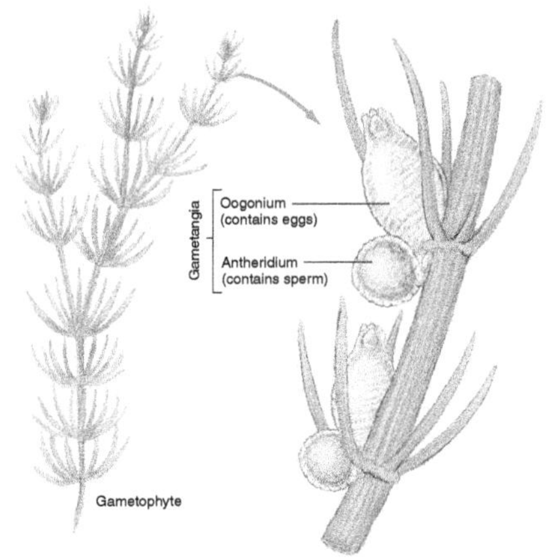

Horsetails

Genus Equisetum

- Only extant genus
- Grows in wet habitats
- Have true roots, stems, and leaves
- Silica is found in the stems making them abrasive
- Used by pioneers to clean pots and pans ("Scouring Rushes")

Club Mosses

Phylum Lycophyta

- Small plants with rhizomes and short erect branches
- Also dominant in the Carboniferous period
- Extant species such as *Lycopodium* are small plants with true roots and stems, and microphylls
- Strobili are found on the stems and produce the spores

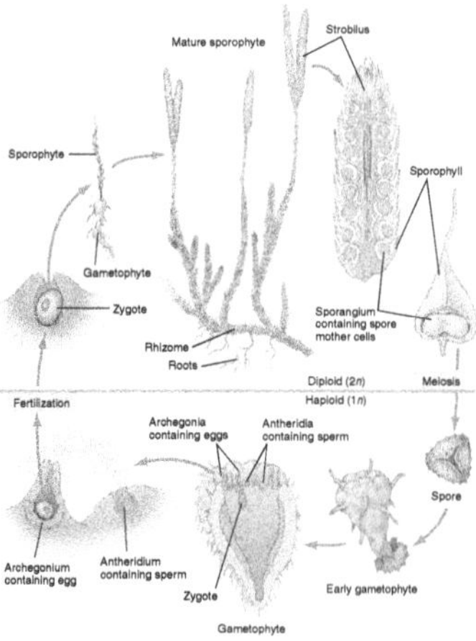

Reproduction
Spore Production

- **Homosporous**—produce only one type of spore
- Examples:
  - Bryophytes
  - Whisk Ferns
  - Horsetails
  - Most Ferns and Club Mosses

Reproduction
Spore Production

- **Heterosporous**—plants that produce Megaspores (Female) and Microspores (Male)
- Was the forerunner of the evolution of seeds
- Examples:
  - Some mosses and ferns
  - Seed plants

Reproduction

Microspore Production

- **Microsporangia**—produce microsporocytes that produce microspores by meiosis
- **Microspores**—can develop into male gametophyte that produces sperm in anteridium

Reproduction

Megaspore Production

- **Megasporangia**—produces megasporocytes which produce megaspores by meiosis
- **Megasporocytes**—can develop into female gametophytes which produce eggs in archegonia

Evolution
Adaptation to Life on Land

- Less dependent on water as a transport medium for reproductive cells
- **Algae**—typically have motile spores and sperm
- **Primitive Plants**—have motile sperm but nonmotile spores
- **Advanced Plants**—have nonmotile spores and sperm

Gymnosperms
"Naked Seed" Plants

- Include some of the largest plants -
  - Giant Sequoia - weighs over 6000 tons
- Tallest plants -
  - Coastal Redwood - 375 feet tall
- Oldest trees -
  - Bristlecone Pines - about 4900 years old

Image Courtesy of Tom Vogel

Gymnosperms
Polyphyletic

- May be classified into four phyla
- Coniferophyta
  - The largest phylum
  - Comprised of about five hundred fifty species
- Ginkophyta and Cycadophyta
  - Evolutionary remnants of once very successful groups
- Gnetophyta
  - An odd group with questionable relationships with the rest of the group

Phylum Coniferophyta
The Conifers

- Woody trees or shrubs
- Do not typically lose their leaves all at once
- Produce resins
- Most are **Monoecious** (bearing both male and female cones on the same tree)
- Dominate in cooler areas of the Northern Hemisphere and some areas of South America, Australia, and Malaysia

Image Courtesy of Tom Vogel

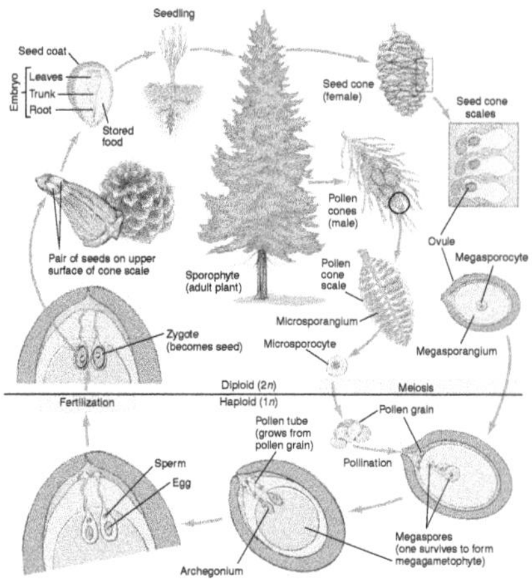

Phylum Coniferophyta
Typical Life Cycle

- The pine tree is the Sporophyte
- Produces Microspores in male cones
- Produces Megaspores in female cones

Phylum Coniferophyta
Typical Life Cycle

- Male Cones
- Have **Microsporophylls** (specialized leaf that bears Microsporangia)
- **Microsporangia**—contain the Microsporocytes
- **Microsporocytes**—undergo meiosis to produce haploid Microspores
- **Microspores**—develop into the reduced male **Gametophyte (Pollen Grain)**

Phylum Coniferophyta
Typical Life Cycle

- Female Cones
  - Are the typical pine cones
  - Larger than the male cones
  - Found on the upper branches of the tree
- Contain Megasporangia that contain Megasporocytes
- **Megasporocytes**—undergo meiosis to produce four haploid Megaspores
- **Megaspores**—one of which will develop into the female gametophyte (the others degenerate)

Typical Life Cycle
Pollination

- Female Cone
  - Opens at the time the pollen is released
  - Pollen grains fall between the scales
  - Stick to a drop of fluid
  - As the fluid evaporates:
    - Pollen is drawn closer to the megasporangium
    - The scales close

Typical Life Cycle
Fertilization

- The cell within the pollen grain divides
- Forms two nonflagellated sperm
- These travel down the pollen tube
- Enter into the female gametophyte
  - One sperm fuses with the egg to form the zygote
  - The other sperm degenerates

Typical Life Cycle
Development

- The tissue of the female gametophyte becomes nutritive tissue in the mature seed
- A seed coat develops around the nutritive tissue and the embryo
- A papery wing develops to allow dispersal by the wind

Typical Life Cycle
Development

- Pollination occurs in Spring
- Due to the slow growth of the pollen tube, fertilization occurs the following year
- Seed development takes several months
- Seeds may remain within the cone for several years before they shed

Typical Life Cycle
Alternation of Generations

- **Sporophyte** is the dominant form
  - Diploid
  - Spore-producing plant
  - Arises from the zygote
- **Gametophyte** is very reduced (microscopic structures within the cones)
  - Organisms arising from the products of meiotic division (spores)
  - Produce the gametes

Typical Life Cycle
Alternation of Generations

- **Archegonia** -
  - Produced by the female gametophyte
  - Multicellular female **Gametangium**
  - Location of the female gamete (egg)
- **Antheridia** -
  - Not produced by the male gametophyte
  - Single or multicellular male gametangium

Cycadophyta

- Were abundant during the Triassic period of the Mesozoic era ("Age of Cycads")
- Only about hundred species remain today
- Many of these are endanger because they are over-collected and sold as ornamental plants
- Restricted to the tropics and subtropics
- Have seed cones and compound leaves

Cycadophyta

- Leaves resemble ferns or palms
- Reproduction is similar to pines
- EXCEPT -
  - Dioecious (cones are on separate male and female plants)
  - Have retained the primitive feature of flagellated sperm
  - Develop a pollen tube, so water is not needed for fertilization

Ginkgophyta

- Represented by a single extant species, *Ginkgo biloba*
- Deciduous
- Dioecious
- Produces flagellated sperm
- Develops a pollen tube
- Used medicinally for centuries

Ginkgophyta

- Extinct in the wild
  - All trees today come from cultivated species in China
- Commonly planted in cities, as they are resistant to air pollution
  - Some cities have prohibitions on planting female trees, because of the odorous fruit they produce

Gnetophytes

Three Unusual Genera

- Contain seventy species
- Three Genera
- Xylem cells resemble the vessel elements of angiosperms

Gnetophytes

- *Gnetum*—a genus of tropical vines and trees with characteristics of flowering plants
- *Ephedra*—is a shrub found in dry regions that resemble "Horsetails"
  - Ephedrine-used to treat asthma
  - Has been involved in deaths due to overdoses when sold as unregulated herbal medicine
- *Welwitschia*—monospecific genus found in southwestern African deserts
  - With only two leaves that shred over time

UNIT FIVE

UNIT V

Ecology
Ecosystems
Climate
Man's Impact

Plant Ecology

Ecological Succession

- A change in a community composition over time
  - **Primary Succession**—occurs in areas where there is no soil formation, such as after a volcanic eruption or glacial retreat
  - **Secondary Succession**—occurs in areas where soil is present, such as when a cultivated field returns to its natural state

Ecological Succession

Models of Succession

- **Climax-Pattern Model**—states that particular areas will always lead to the same type of community "Climax Community"
- Climate is one of the determining factors of which plant community will be established
  - Others include:
    - Altitude
    - Soil type
    - Animal interaction

Ecological Succession
Models of Succession

- **Facilitation Model**—states that each stage in succession facilitates the next stage
- Example:
  - Shrubs cannot grow on sand dunes until dune grasses make the soil suitable for them

Ecological Succession
Models of Succession

- **Inhibition Model**—states that colonists hold on to their space and inhibit the growth of other plants until the colonists die or are damaged
- **Tolerance Model**—states that different kinds of plants can colonize an area at the same time

Intermediate Disturbance Hypothesis

- States that a moderate amount of disturbance is required for a high degree of community diversity
- Examples:
  - Fire
  - Flood

Island Biogeography

- States that every island has a balance between the rate of immigration of new species and the rate of extinction of species that are already on the island
  - No two species will occupy the same niche
  - The "new" species will not move in until the "old" species is no longer the "Fittest"

Biosphere
Climate

- The distribution of Biomes in the Biosphere depends on:
  1. Variations in reception of solar radiation due to a spherical earth
  2. The tilt of the earth's axis as it rotates about the sun
  3. Distribution of land masses and oceans
  4. Topography (Landscape) features

Air Circulation

- The tilt of the earth causes an unequal heating of the earth's surface creating the seasonal changes, except at the Equator
- When the Northern Hemisphere has winter, the Southern Hemisphere has summer
- Heat always passes from warmer areas to cooler areas

Air Circulation

- Because the Earth rotates, these moving air masses (Winds) are modified into three large circulation cells in each hemisphere
- At the Equator, the Sun heats the air and evaporates the water
- The warm moist air rises and cools, losing much of its moisture as rain
- The greatest amounts of rainfall are near the Equator

Air Circulation

Topography

- Means the physical features of the land
- One topographical feature that affects the climate is the presence of mountains
    - As air blows up and over, the mountain range rises and cools
    - This side of the mountain (windward) receives much more rain than the other (leeward) side
- The leeward side of the mountain is said to be in a "Rain Shadow"

Biomes of the World

- The largest geographical unit of the biosphere
- Has a particular mix of plants and animals that have adapted to living under certain environmental conditions
- Biomes gradually change from one type to another

Terrestrial Biomes
Arctic Tundra

- Encircles the Earth just south of the ice-covered polar seas in the Northern Hemisphere
  - Covers about 20% of the earth's land surface
  - Cold and dark much of the year
  - Only about 20 cm of rainfall per year
  - Only the topmost layer of earth ever thaws, below that is always frozen (Permafrost)
  - No trees (growing season is too short, and permafrost prevents adequate root growth)
  - Few animals live in the Tundra year round

Terrestrial Biomes
Coniferous Forests

- Found in three locations:
- **Taiga**—extends around the world in the northern part of North America and Eurasia
- Near mountain tops (Montane Coniferous Forest)
- Along the Pacific coast of North America as far south as California

Terrestrial Biomes
Coniferous Forests

- **Taiga (Swampland)**—typifies the coniferous forest with its cone-bearing trees such as spruce, fir, and pine that do not lose their leaves seasonally
  - Well adapted to the cold with thick coverings on both leaves and bark
  - Leaves (needles) are modified to prevent water loss, freezing, and damage due to heavy snow
  - Limited understory growth

Terrestrial Biomes
Temperate Deciduous Forest

- Found south of the Taiga in North America, eastern Asia, and much of Europe
- Moderate climate with relatively high rainfall (75–150 cm)
- Well-defined seasons, with a 140–300 day growing season
- Trees such as oak, beech, maple, hickory, and others with broad, flat leaves
- Grow leaves in the Spring and lose them in the Fall

Terrestrial Biomes
Tropical Forests

- Found in South America, Africa, and the Indo- Malayan region near the Equator
- The weather is always warm, and rainfall is plentiful (Minimum of 190 cm)
- May be the richest terrestrial biome both in numbers of different species and their abundance

Terrestrial Biomes
Tropical Forests

- Divided into three levels:
  - Forest Floor
  - Understory
  - Canopy

Terrestrial Biomes
Grasslands

- Occur where rainfall is greater than 25 cm
  - generally too low to support trees
- Once covered more than 40% of the earth's surface (Many areas are now agricultural)
- Grasses are well adapted to a changing environment
- Can tolerate overgrazing, flood, drought, and even fire

Terrestrial Biomes
Prairies

- Can be divided into two types:
- **Tall Grass Prairies**
  - Requires more rainfall
  - Large tall grasses (>2 m)
- **Short Grass Prairies**
  - Require less rainfall
  - Small, short grasses between 5 and 10 cm

Terrestrial Biomes
Savannas

- Grasslands that contain some trees
- Occur in regions where a relatively cool, dry season is followed by a hot, rainy season
- Supports a large variety of grazers

Terrestrial Biomes

Desert

- Usually found at about 30° latitudes in the Northern and Southern Hemispheres
- Winds in these regions lack moisture
- Annual rainfall is less than 25 cm
- Days are hot, and nights are cold

Food, Soil, and Pest Management

Food Production Has Increased Dramatically

- About 10,000 years ago, humans began to shift from hunting for and gathering their food to growing it and raising animals for food and labor.
- Today, three systems supply most of our food.
  - Croplands produce mostly grains.
  - Rangelands, pastures, and feedlots produce meat.
  - Fisheries and aquaculture provide us with seafood.
- About 66% of the world's people survive primarily by eating rice, wheat, and corn.
  - Only a few species of mammals and fish provide most of the world's meat and seafood.

Food Production Has Increased Dramatically

- Since 1960, there has been an increase in global food production from all three of the major food production systems because of technological advances.
  - Tractors, farm machinery, and high-tech fishing equipment.
  - Irrigation.
  - Inorganic chemical fertilizers, pesticides, high-yield grain varieties, and industrialized production of livestock and fish

Industrialized Crop Production Relies on High-Input Monocultures

- Agriculture used to grow crops can be divided roughly into two types:
  - Industrialized agriculture, or high-input agriculture, uses heavy equipment and large amounts of financial capital, fossil fuel, water, commercial inorganic fertilizers, and pesticides to produce single crops, or monocultures.

Industrialized Crop Production Relies on High-Input Monocultures

- Plantation agriculture is a form of industrialized agriculture used primarily in tropical less developed countries.
  - Grows cash crops such as bananas, soybeans, sugarcane, coffee, palm oil, and vegetables.
  - Crops are grown on large monoculture plantations, mostly for export to more developed countries.

Traditional Agriculture Often Relies on Low-Input Polycultures

- Traditional agriculture provides about 20% of the world's food crops on about 75% of its cultivated land, mostly in less developed countries.

Traditional Agriculture Often Relies on Low-Input Polycultures

- Many traditional farmers grow several crops on the same plot simultaneously, a practice known as **polyculture**.
  - Crop diversity reduces the chance of losing most or all of the year's food supply to pests, bad weather, and other misfortunes.
  - Crops mature at different times, provide food throughout the year, reduce the input of human labor, and keep the soil covered to reduce erosion from wind and water.

A Closer Look at Industrialized Crop Production

- Farmers can produce more food by increasing their land or their yields per acre.
- Since 1950, about 88% of the increase in global food production has come from using high-input industrialized agriculture to increase yields in a process called the **Green Revolution**.
- Three steps of the Green Revolution:
  - First, develop and plant monocultures of selectively bred or genetically engineered high-yield varieties of key crops such as rice, wheat, and corn.

A Closer Look at Industrialized Crop Production

  - Second, produce high yields by using large inputs of water and synthetic inorganic fertilizers, and pesticides.
  - Third, increase the number of crops grown per year on a plot of land through multiple cropping.
- The first green revolution used high-input agriculture to dramatically increase crop yields in most of the world's more developed countries, especially the United States, between 1950 and 1970.

A Closer Look at Industrialized Crop Production

- A second Green Revolution has been taking place since 1967. Fast-growing varieties of rice and wheat, specially bred for tropical and subtropical climates, have been introduced into middle-income, less developed countries, such as India, China, and Brazil.

A Closer Look at Industrialized Crop Production

- Largely because of the two Green Revolutions, world grain production tripled between 1961 and 2009.
- People directly consume about 48% of the world's grain production. About 35% is used to feed livestock and indirectly consumed by people who eat meat and meat products. The remaining 17% (mostly corn) is used to make biofuels such as ethanol for cars and other vehicles.

A Closer Look at Industrialized Crop Production

- In the United States, industrialized farming has evolved into agribusiness, as a small number of giant multinational corporations increasingly control the growing, processing, distribution, and sale of food in US and global markets.
- Since 1950, US industrialized agriculture has more than doubled the yields of key crops such as wheat, corn, and soybeans without cultivating more land.
- Americans spend only about 13% of their disposable income on food, compared to the percentages up to 50% that people in China and India and most other less developed countries have to pay for food.

Crossbreeding and Genetic Engineering Produce Varieties of Crops and Livestock

- Crossbreeding through artificial selection has been used for centuries by farmers and scientists to develop genetically improved varieties of crops and livestock animals.

Crossbreeding and Genetic Engineering Produce Varieties of Crops and Livestock

- Modern scientists are creating a second gene revolution by using genetic engineering to develop genetically improved strains of crops and livestock.

Crossbreeding and Genetic Engineering Produce Varieties of Crops and Livestock

- Currently, at least 70% of the food products on US supermarket shelves contain some form of genetically engineered food or ingredients, but no law requires the labeling of genetically modified (GM) products.
- Certified organic food, which is labeled as makes no use of GM seeds or ingredients.

Producing Food Has Major Environmental Impacts

- Spectacular increases in the world's food production since 1950. The bad news is the harmful environmental effects associated with such production increases.
- According to many analysts, agriculture has a greater total harmful environmental impact than any human activity.
- These environmental effects may limit future food production and make it unsustainable.

Topsoil Erosion Is a Serious Problem in Parts of the World

- Soil erosion is the movement of soil components, especially surface litter and topsoil, from one place to another by the actions of wind and water.

Drought and Human Activities Are Degrading Drylands

- Desertification in arid and semiarid parts of the world threatens livestock and crop contributions to the world's food supply.
- Desertification occurs when the productive potential of topsoil falls by 10% or more because of a combination of prolonged drought and human activities that expose topsoil to erosion.

Excessive Irrigation Has Serious Consequences

- Irrigation boosts productivity of farms; roughly 20% of the world's cropland that is irrigated produces about 45% of the world's food.

Excessive Irrigation Has Serious Consequences

- Severe salinization has reduced yields on at least 10% of the world's irrigated cropland and almost 25% of irrigated cropland in the United States, especially in Western states.
- Excessive irrigation contributes to depletion of groundwater and surface water supplies.

Food and Biofuel Production Systems Have Caused Major Losses of Biodiversity

- Natural biodiversity and some ecological services are threatened when forests are cleared and grasslands are plowed up and replaced with croplands used to produce food or biofuels, such as ethanol.
- There is increasing loss of agrobiodiversity, the world's genetic variety of animal and plant species.
- In the United States, about 97% of the food plant varieties that were available to farmers in the 1940s no longer exist, except perhaps in small amounts in seed banks and in the backyards of a few gardeners.

There Is Controversy over Genetically Engineered Foods

- Controversy has arisen over the use of GM food and other products of genetic engineering.
- Its producers and investors thought GM food as a potentially sustainable way to solve world hunger problems and improve human health.

Nature Controls the Populations of Most Pests

- A pest is any species that interferes with human welfare by competing with us for food, invading homes, lawns and gardens, destroying building materials, spreading disease, invading ecosystems, or simply being a nuisance.
- Worldwide, only about hundred species of plants ("weeds"), animals (mostly insects), fungi, and microbes cause most of the damage to the crops we grow.

Shift to More Sustainable Food Production

- Industrialized agriculture produces large amounts of food at reasonable prices, but is unsustainable because it:
  - Relies heavily on fossil fuels
  - Reduces biodiversity and agrobiodiversity
  - Reduces the recycling of plant nutrients back to topsoil

Climate Change

Climate Change

- Climate change occurs when the patterns change in time (e.g., winter months get warmer) and space (e.g., monsoon rains occur further south).
- Climate changes naturally on a range of timescales, from decadal (10's of years), centennial (100's of years), millennial (1000's of years), and longer (glacial cycles, e.g., Ice Ages).
- Climate changes naturally on a range of spatial scales, from local and regional to global.
- Climate determines the type and location of human-managed ecosystems, such as agricultural farmlands.
- Climate affects the weathering of rock, the type of soil that forms, and the rate of soil formation.

- Climate largely determines the nature and locations of **biomes** (major terrestrial ecosystems, defined based on their plant communities).
- Example: a prediction of climate change for eastern PA. By the end of this century, Philadelphia, PA, could have the climate that Savanna, GA, has now.

Air bubbles in ice cores retain atmospheric gases present when the ice was formed

CO_2 measured continuously at Mauna Loa Observatory since 1958 has provided strong evidence for atmospheric change

Annual tree rings not only indicate tree age but also the conditions under which the tree grew

- Energy retention in the atmosphere depends on the abundance and effectiveness of the Greenhouse Gas (GHG).
- The diagram shows how much each gas contributed to warming from 1750 to 2005 (termed **radiative forcing**)

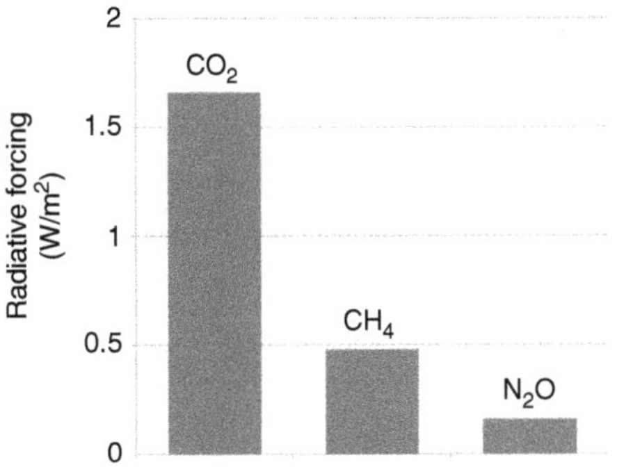

- Only about 50% of the increased CO_2 stays in the atmosphere.
- The rest is absorbed by the oceans and other sinks.
- CO_2 dissolved in the oceans is forming extra carbonic acid, causing a decline in pH and other ecosystem problems

- Consequences of global climate change include:
  - Melting of glaciers and sea ice
  - Rising sea level
  - Changes to the ecosystems and survival of natural species
  - Changes to agricultural conditions reducing our food supply
  - Increased threats to human health from new diseases and changing disease distributions

Glaciers are Melting

According to the National Snow and Ice Data Center:

"Because glaciers are sensitive to temperature fluctuations, they provide clues about the effects of global warming. With few exceptions, glaciers around the world have retreated at unprecedented rates over the last century. Some ice caps, glaciers, and even an ice shelf have disappeared altogether. Many more are retreating so rapidly that they may vanish within decades."

Melting glaciers supply drinking and irrigation water for hundreds of millions of people around the world.

In addition, melting glaciers lead to rising sea levels drying regional climate.

- Because most glaciers are melting, sea level is rising.
- In addition, the oceans' surfaces are warming because of climate change, thus causing expansion of the water and adds to sea-level rise.
- From 1993 to 2003, the estimated rate of sea-level rise from both causes has been about 2.4 cm/decade.
  - Oceans are predicted to rise about 12 cm in 50 years.
- Coastal erosion rates are about 100 times the rate of sea-level rise, meaning that every centimeter of sea-level rise results in 1 m of beach erosion.

The Implications of Melting Sea Ice

- Sea ice tends to reflect sun's energy; sea water tends to absorb sun's energy.
- When more sea ice melts in summer, there is more open water to absorb energy and heat up.
- The following winter, it takes longer for sea ice to form due to warmer water and thinner ice.
- Then, more melting takes place the next summer, which warms the water even more . . . and so on.
- The ". . . and so on" is an indicator of a **positive feedback process**. Now that melting has started, it might accelerate because of the positive feedback.

Melting Permafrost

- Warming climate is melting the permafrost in Canada, Alaska, and Scandinavia.
- This increases wetlands.
- The wetlands release methane and carbon dioxide, which are GHGs. . ., thus more **positive feedback effects**.

Practices and Technologies

We can reduce emissions by:

- **Conservation**: use less energy (e.g., turn off lights, machines, and engines when they're not in use)
- **Conversion**: change to low-emission energy sources (e.g., renewable fuels emit fewer GHG than fossil fuels)
- **Efficiency**: use technologies that get the most from an energy source (e.g., Energy Star appliances)
- **Waste reduction**: eliminate materials that require energy to landfill (e.g., compost organics instead of creating trash)
- **Sequestration**: take GHG out of the air (e.g., maintain and increase crops and land uses that hold carbon in soils)

What Can You Do?

What can you do personally?

The most important step is to find out how much GHG your lifestyle creates. Use one or more of the following online calculators to find out. Then use the GHG-reducing tips on these sites to reduce your GHG production.

The US EPA household emission calculator (http://www.epa.gov/climatechange/emissions/ind_calculator.html)

The Nature Conservancy carbon footprint calculator (http://www.nature.org/initiatives/climatechange/calculator/ Carbonfootprint.com

(http://www.carbonfootprint.com/)

Water Resources and Water Pollution

Freshwater Is an Irreplaceable Resource That We Are Managing Poorly

- Freshwater is relatively pure and contains few dissolved salts.
- Earth has a precious layer of water—most of it saltwater—covering about 71% of the earth's surface.
- Water is an irreplaceable chemical with unique properties that keep us and other forms of life alive.

Freshwater Is an Irreplaceable Resource That We Are Managing Poorly

- Concerns regarding water include:
  - Access to freshwater is a global health issue. Every day an average of 3900 children younger than 5 years die from waterborne infectious diseases.
  - An economic issue—vital for reducing poverty and producing food and energy.
  - A women's and children's issue in developing countries because poor women and girls are often responsible for finding and carrying daily supplies of water.

Most of the Earth's Freshwater Is Not Available to Us

- About 0.024% is readily available to us as liquid freshwater in accessible groundwater deposits and in lakes, rivers, and streams.
- The rest is in the salty oceans, in frozen polar ice caps and glaciers, or in deep underground and inaccessible locations.

Most of the Earth's Freshwater Is Not Available to Us

- Freshwater is not distributed evenly.
  - Differences in average annual precipitation and economic resources divide the world's continents, countries, and people into water haves and have-nots.
  - Canada, with only 0.5% of the world's population, has 20% of the world's liquid freshwater, while China, with 19% of the world's people, has only 7% of the supply.

Groundwater and Surface Water Are Critical Resources

- Surface water is the freshwater from precipitation and snowmelt that flows across the earth's land surface and into lakes, wetlands, streams, rivers, estuaries, and ultimately to the oceans.
  - Precipitation that does not infiltrate the ground or return to the atmosphere by evaporation is called **surface runoff**.
  - The land from which surface water drains into a particular river, lake, wetland, or other body of water is called its **watershed**, or **drainage basin**.

We Use a Large and Growing Portion of the World's Reliable Runoff

- Two-thirds of the annual surface runoff in rivers and streams is lost by seasonal floods and is not available for human use.
  - The remaining one third is reliable surface runoff, which we can generally count on as a source of freshwater from year to year.
- During the past century, the human population tripled, global water withdrawals increased sevenfold, and per capita withdrawals quadrupled. We now withdraw about 34% of the world's reliable runoff of freshwater.

We Use a Large and Growing Portion of the World's Reliable Runoff

- Worldwide, about 70% of the water we withdraw each year comes from rivers, lakes, and aquifers to irrigate cropland, industry uses another 20%, and residences 10%.
- Affluent lifestyles require large amounts of water.

Freshwater Shortages Will Grow

- More than thirty countries—mainly in the Middle East and Africa—now face water scarcity.
- By 2050, sixty countries, many of them in Asia, with three-fourths of the world's population, are likely to be suffering from water stress.

Freshwater Shortages Will Grow

- In 2009, about 1 billion people in the world currently lack regular access to enough clean water for drinking, cooking, and washing.
- By 2025, at least 3 billion people are likely to lack access to clean water.

Groundwater Is Being Withdrawn Faster Than It Is Replenished in Some Areas

- Aquifers provide drinking water for nearly half of the world's people.
- Most aquifers are renewable resources unless their water becomes contaminated or is removed faster than it is replenished by rainfall.
- Water tables are falling in many areas of the world because the rate of pumping water from aquifers (mostly to irrigate crops) exceeds the rate of natural recharge from rainfall and snowmelt.

Overpumping of Aquifers Has Several Harmful Effects

- As water tables drop, farmers must drill deeper wells, buy larger pumps, and use more electricity to run those pumps. Poor farmers cannot afford to do this and end up losing their land.
- Withdrawing large amounts of groundwater causes the sand and rock in aquifers to collapse.

Large Dams and Reservoirs Have Advantages and Disadvantages

- Dams are structures built across rivers to block some of the flow of water.
- Dammed water usually creates a reservoir, a store of water collected behind the dam.
- A dam and reservoir:
  - capture and store runoff and release it as needed to control floods.
  - generate electricity (hydroelectricity)

Large Dams and Reservoirs Have Advantages and Disadvantages

  - supply water for irrigation and for towns and cities.
  - provide recreational activities such as swimming, fishing, and boating.
- The world's 45,000 large dams have increased the annual reliable runoff available for human use by nearly 33%.
- Negative effects of dams include:
  - displaced 40–80 million people from their homes.

Large Dams and Reservoirs Have Advantages and Disadvantages

    - flooded an area of mostly productive land totaling roughly the area of California.
    - impaired some of the important ecological services that rivers provide.
- Reservoirs eventually fill up with sediment, typically within 50 years, eventually making them useless for storing water or producing electricity.
- Around five hundred small dams have been removed in the US, but removal of large dams is controversial and expensive.

Reducing Freshwater Waste Has Many Benefits

- An estimated 66% of the freshwater used in the world is unnecessarily wasted.
- In the United States—the world's largest user of water—about half of the water drawn from surface and groundwater supplies is wasted.
- It is economically and technically feasible to reduce such water losses to 15%, thereby meeting most of the world's water needs for the foreseeable future.

Reducing Freshwater Waste Has Many Benefits

- Reasons so much freshwater is wasted:
  - Government subsidies that keep the cost of freshwater low.
  - Lack of government subsidies for improving the efficiency of freshwater use.

We Can Cut Freshwater Waste in Irrigation

- About 60% of the irrigation water worldwide does not reach the targeted crops.
- In most irrigation systems, water is pumped from a groundwater or surface water source through unlined ditches, and about 40% is lost through evaporation, seepage, and runoff.
- Flood irrigation delivers far more water than needed for crop growth and typically loses 40% of water through evaporation, seepage, and runoff.
- With existing irrigation, could be reduced to 5%–10%.

We Can Cut Freshwater Waste in Industry and Homes

- Producers of chemicals, paper, oil, coal, primary metals, and processed food consume almost 90% of the water used by industry in the United States.
- Some of these industries recapture, purify, and recycle water to reduce their water use and water treatment costs.
- Most industrial processes could be redesigned to use much less freshwater.

We Can Cut Freshwater Waste in Industry and Homes

- Flushing toilets with freshwater is the largest use of domestic water in the United States.
  - Standards have required that new toilets use no more than 6.1 L (1.6 gallons) of water per flush.
- Studies show that 30%–60% of the freshwater supplied in nearly all of the world's major cities in less developed countries is lost, primarily through leakage in water mains, pipes, pumps, and valves.

We Can Cut Freshwater Waste in Industry and Homes

- Fixing leaks should be a high government priority, would cost less than building dams or importing water.
- Homeowners and businesses in water-shortage areas are using drip irrigation and replacing lawns with native plants that need little freshwater.
- About 50%–75% of the slightly dirtied water from bathtubs, showers, sinks, dishwashers, and clothes washers in a typical house could be stored in a holding tank and then reused as gray water to irrigate lawns and nonedible plants, to flush toilets, and to wash cars.

We Can Cut Freshwater Waste in Industry and Homes

- The relatively low cost of water in most communities causes excessive water use and waste.
  - Many water utility and irrigation authorities charge a flat fee for water use, and some charge less for the largest users of water.
  - About one-fifth of all US public water systems do not have water meters and charge a single low rate for almost unlimited use of high-quality water.
  - Many apartment dwellers have little incentive to conserve water, because water use charges are included in their rent.

We Can Use Less Water to Remove Wastes

- Large amounts of freshwater good enough to drink are being flushed away as industrial, animal, and household wastes.
- Within 40 years, we may need the world's entire reliable flow of river water just to dilute and transport the wastes we produce each year.
- Save water by using systems that mimic the way nature deals with wastes by recycling them.
- Rely more on waterless composting toilets.

Major Water Pollutants Have Harmful Effects

- According to the World Health Organization, an estimated 4400 people die each day from preventable infectious diseases that they got infected from drinking contaminated water.

Reducing Surface Water Pollution from Nonpoint Sources

- There are a number of ways to reduce nonpoint-source water pollution, most of which comes from agriculture.
  - Reduce soil erosion by keeping cropland covered with vegetation.
  - Reduce the amount of fertilizer that runs off into surface waters and leaches into aquifers by using slow-release fertilizer, using no fertilizer on steeply sloped land, and planting buffer zones of vegetation between cultivated fields and nearby surface waters.

Reducing Surface Water Pollution from Nonpoint Sources

- Organic farming can also help prevent water pollution caused by nutrient overload.
- Control runoff and infiltration of manure from animal feedlots by planting buffers and locating feedlots and animal waste sites away from steeply sloped land, surface water, and flood zones.

Laws Can Help to Reduce Water Pollution From Point Sources

- The Federal Water Pollution Control Act of 1972 (renamed the Clean Water Act) and the 1987 Water Quality Act form the basis of US efforts to control pollution of the country's surface waters.
  - Set standards for allowed levels of key water pollutants and require polluters to get permits limiting how much of various pollutants they can discharge into aquatic systems.

Laws Can Help to Reduce Water Pollution From Point Sources

- According to the Environmental Protection Agency (EPA), the Clean Water Act of 1972 led to numerous improvements in US water quality. Between 1992 and 2002:
  - The percentage of US stream lengths found to be fishable and swimmable increased from 36% to 60% of those tested.
  - The proportion of the US population served by sewage treatment plants increased from 32% to 74%.
  - Annual wetland losses decreased by 80%.

Laws Can Help to Reduce Water Pollution from Point Sources

- More work to be done:
  - In 2006, the EPA found that 45% of the country's lakes and 40% of streams surveyed were still too polluted for swimming or fishing, and that runoff of animal wastes from feedlots and meat processing facilities pollutes seven of every ten US rivers.
  - Fish caught in more than 1400 different waterways and more than one-fourth of the nation's lakes are unsafe to eat because of high levels of pesticides, mercury, and other toxic substances.
  - A 2007 government study found that tens of thousands of gasoline storage tanks in in forty-three states states are leaking.

www.ingramcontent.com/pod-product-compliance
Ingram Content Group UK Ltd.
Pitfield, Milton Keynes, MK11 3LW, UK
UKHW051333111026
15762UKWH00030B/869

9 781524 968915